Dictionary of Statistics for Psychologists

B. S. EVERITT *and* **T. WYKES**

Institute of Psychiatry, University of London

A member of the Hodder Headline Group
LONDON • SYDNEY • AUCKLAND
Co-published in the United States of America by
Oxford University Press Inc., New York

First published in Great Britain in 1999 by
Arnold, a member of the Hodder Headline Group,
338 Euston Road, London NW1 3BH

http://www.arnoldpublishers.com

Co-published in the United States of America by
Oxford University Press Inc.,
198 Madison Avenue, New York, NY 10016

British Library Cataloguing in Publication Data
A catalogue record for this book is available from the British Library

Library of Congress Cataloging-in-Publication Data
A catalog record for this book is available from the Library of Congress

ISBN 0 340 71997 4 (hb)
ISBN 0 340 71998 2 (pb)

1 2 3 4 5 6 7 8 9 10

Publisher: Nicki Dennis
Production Editor: Liz Gooster
Production Controller: Iain McWilliams
Cover Design: Terry Griffiths

Typeset in 9/11 pt Times by Academic & Technical, Bristol
Printed and bound in Great Britain by MPG Books, Bodmin, Cornwall

Preface

Psychologists need to know about statistics and there is a vast variety of excellent statistics textbooks aimed at them. But few, if any, of the available texts cover the whole range of statistical methodology used in psychological research and practice, particularly in the relatively concise terms often required by psychology students and working psychologists. This dictionary aims to fill this gap and provides definitions of over 1500 statistical terms commonly encountered in the psychological literature. In many cases some mathematical details and a numerical example follow a brief written account of a term. In addition graphical material is given for many entries. We hope that this combination will be successful in making the definition clear to most readers. For many topics a reference to the term's use in a psychological journal or book is also included.

Various forms of cross-referencing are used. Terms in **Comic Sans MS** in an entry appear as a separate headword, although commonly occurring terms – for example, variable, population, sample, mean, significance level, degrees of freedom, explanatory variable, etc. – are not referred to in this way. Some entries simply refer readers to another entry. This may indicate that the terms are synonymous or, alternatively, that the term is more conveniently discussed under another entry. In the latter case the term is printed in *italics* in the entry where it is discussed most fully. Entries are in alphabetical order using the letter-by-letter rather than the word-by-word convention. In terms containing numbers or Greek letters, the numbers or corresponding English word are spelt out and alphabetized accordingly. So, for example, 2×2 table is found under **two-by-two table**, and α-trimmed mean under **alpha-trimmed mean**. No headings are inverted: for example, there is an entry under **end-aversion bias** *not* under **bias, end-aversion**.

Acknowledgements

First we would like to thank the many authors who have, unwittingly, provided the basis of a large number of the definitions included in this dictionary through their books and papers. Next our enormous gratitude is due to Mrs Harriet Meteyard for maintaining and typing the files of the many versions that preceded the final manuscript. Finally a special thank you to Bethan (9 years) and Megan (10 years) Davies Wykes, for checking the cross-referencing. Any errors that remain are entirely due to them, and where noticed by readers, will lead to a radical reduction in pocket money.

Acceptance region The set of values of a **test statistic** for which the null hypothesis is accepted. Suppose, for example, a **z-test** is being used to test that the mean of a population is 10 against the alternative hypothesis that it is not 10. If the significance level chosen is 0.05 then the acceptance region consists of values of z between -1.96 and 1.96.

Accidentally empty cells Synonym for **sampling zeros**.

Accuracy The degree of conformity to some recognized standard value. See also **bias**.

ACF Abbreviation for **autocorrelation function**.

Acquiescence bias The **bias** produced by respondents in a **sample survey** who have the tendency to give positive responses, such as 'true', 'like', 'often' or 'yes' to a question. At its most extreme, the person responds in this way irrespective of the content of the item. Thus a person may respond 'true' to two items like 'I always arrive at appointments on time' and 'I am often late for appointments'. See also **end-aversion bias**. [*Journal of Community and Applied Social Psychology*, 1996, 6, 207–227.]

Actuarial statistics The statistics used by actuaries to evaluate risks, calculate liabilities and plan the financial course of insurance, pensions etc. An example is **life expectancy** for people of various ages, occupations, etc.

Adaptive methods Procedures that use various aspects of the sample data to select the most appropriate type of statistical method for analysis. An adaptive estimator, T, for the centre of a **frequency distribution**, for example, might be:

$$\begin{aligned} T &= \text{mid range when } k \leq 2 \\ &= \text{arithmetic mean when } 2 < k < 5 \\ &= \text{median when } k \geq 5 \end{aligned}$$

where k is the sample **skewness** of the distribution. So if the sample looks as if it arises from a short-tailed distribution, the average of the largest and smallest observations is used; if it looks like a long-tailed situation the median is used; otherwise the mean of the sample is employed.

Addition rule for probabilities For two **mutually exclusive events**, A and B, the probability of either event occurring is the sum of the separate probabilities.

Mathematical details

The addition rule states that:

$$\Pr(A \text{ or } B) = \Pr(A) + \Pr(B)$$

where $\Pr(A)$ denotes the probability of event A etc. For k mutually exclusive events $A_1, A_2, \ldots, A_k$ the more general rule is

$$\Pr(A_1 \text{ or } A_2 \cdots \text{ or } A_k) = \Pr(A_1) + \Pr(A_2) + \cdots + \Pr(A_k)$$

Numerical example

If a population consists of 60% English people, 20% Scottish, 10% Welsh and 10% Irish, then the probability that an individual selected at random from the population is a Celt is

$$\Pr(\text{Celt}) = \Pr(\text{Scottish}) + \Pr(\text{Welsh}) + \Pr(\text{Irish}) = 0.20 + 0.10 + 0.10 = 0.40$$

See also **multiplication rule for probabilities**.

Additive effect A term used when the effect of administering two factors together is the sum of their separate effects. So, for example, if the effect of factor A is a units and of factor B is b units, the effect of A and B together is $a + b$ units. See also **additive model**.

Additive model A model in which the explanatory variables have an **additive effect** on the dependent variable. So, for example, if variable A has an effect of size a on some dependent measure and variable B one of size b on the same response, then in an assumed additive model for A and B their combined effect would be $a + b$.

Age heaping A term applied to the collection of data on ages when these are accurate only to the nearest year, half year or month. See also **rounding**.

Agglomerative hierarchical clustering methods Methods of **cluster analysis** that begin with each individual in a separate cluster and then, in a series of steps, combine individuals and, later, clusters into new, larger clusters until a final stage is reached where all individuals are members of a single group. At each stage the individuals or clusters that are 'closest', according to some particular definition of distance, are joined. The whole process can be summarized by a **dendrogram**. Solutions corresponding to particular numbers of clusters are found by 'cutting' the dendrogram at the appropriate level. See also **complete linkage cluster analysis**, **single linkage clustering**, **Ward's method**, and **K-means cluster analysis**.

AID Abbreviation for **automatic interaction detector**. [*Mental Health in Black America.* Neighbors, Harold W (Ed); Jackson, James Sidney (Ed); et al 1996 (pp 14–26). Thousand Oaks, CA: Sage].

Algorithm A well-defined set of rules which, when routinely applied, leads to a solution of a particular class of mathematical or computational problem.

All subsets regression A form of **regression analysis** involving consideration of all possible regression equations linking the dependent and explanatory variables. The 'best' equation is selected by comparing the values of some chosen numerical criterion calculated for each. A number of such criteria are available of which the most commonly used is **Mallow's C_k statistic**. If there are q explanatory variables then a total of $2^q - 1$ equations need to be examined. So, for example, if $q = 3$ the chosen numerical criterion would be calculated for the following seven regressions:

1. y on x_1
2. y on x_2
3. y on x_3
4. y on x_1 and x_2
5. y on x_1 and x_3
6. y on x_2 and x_3
7. y on x_1, x_2 and x_3

If the C_k statistic were calculated, the best equation would be the one with the lowest value. (A numerical example is given under the **Mallow's C_k statistic** statistic entry.) See also **selection methods in regression**.

Alpha (α) The probability of a **Type I error**. See also **significance level**.

Alternative hypothesis The hypothesis against which the null hypothesis is tested. Often also known as the experimental hypothesis.

Analysis of covariance An extension of the **analysis of variance** that allows for the possible effects of continuous concomitant variables (**covariates**) on the response variable, in addition to the effects of the factor variables. The assumption is usually made that there is a linear relationship between the response variable and the covariates and this relationship is identical in each combination of factor levels. If these assumptions are true then inclusion of covariates in this way decreases the residual mean square of the observations and so increases the sensitivity of the ***F*-tests** used in assessing differences between the levels of each factor and their interactions.

Mathematical details

For a one-factor design with k groups the model for the dependent variable is

$$y_{ij} = \mu + \alpha_i + \beta x_{ij} + \varepsilon_{ij}$$

where y_{ij} is the value of the dependent variable for the jth individual in the ith group, μ is an overall mean, α_i is the group effect, x_{ij} is the value of the covariate for the individual, β is the regression coefficient of the dependent variable on the covariate (assumed the same in all groups) and ε_{ij} is the residual or error term, assumed to have a **normal distribution** with mean zero and variance σ^2.

Numerical example

Twenty subjects were given a behaviour approach test to determine how close they could walk to a snake without feeling uncomfortable. This score was taken as the covariate. Next they were randomly assigned to one of four treatments, the first of which was a control (placebo) while the other three were different methods intended to reduce a human's fear of snakes. After treatment each subject's approach test score was taken again. The data are shown below:

Control		Method 1		Method 2		Method 3	
Initial	Final	Initial	Final	Initial	Final	Initial	Final
25	25	17	11	32	24	10	8
13	25	9	9	30	18	29	17
10	12	19	16	12	2	7	8
25	30	25	17	30	24	17	12
10	37	6	1	10	2	8	7

The results of applying an analysis of covariance to the data are summarized below:

Source	SS	df	MS	*F*	*P*
Groups	806.8	3	268.9	8.08	0.0019
Initial score	573.6	1	573.6	17.24	0.00019
Error	499.2	15	33.3		

The analysis demonstrates both that there is a significant relationship between final score and initial score and that the groups differ with respect to mean final score after controlling for initial score.

Analysis of variance The separation of variance attributable to one cause from the variance attributable to others. By partitioning the total variance of a set of observations into parts due to particular factors – for example, sex, treatment group – and comparing variances by way of **F-tests**, differences between means can be assessed. The simplest analysis of this type involves a *one-way design*, in which N subjects are allocated, usually at random, to the k different levels of a single factor. The total variation in the observations is then divided into a part due to differences between level means (the *between groups sum of squares*) and a part due to the differences between subjects in the same group (the *within groups sum of squares*, also known as the *residual sum of squares*). These terms are usually arranged as an *analysis of variance table*.

Source	df	SS	MS	MSR
Between groups	$k-1$	SSB	$\mathrm{SSB}/(k-1)$	$\dfrac{\mathrm{SSB}/(k-1)}{\mathrm{SSW}/(N-K)}$
Within groups	$N-k$	SSW	$\mathrm{SSW}/(N-k)$	
Total	$N-1$			

SS = sum of squares; MS = mean square; MSR = mean square ratio

If the means of the populations represented by the factor levels are the same, then the *between groups mean square* and *within groups mean square* are both estimates of the same population variance. Whether this is so can be assessed by a suitable F-test on the mean square ratio.

Mathematical details

Model for one-way analysis of variance:

$$y_{ij} = \mu + \alpha_i + \varepsilon_{ij}$$

where y_{ij} is the jth score in group i, μ is the overall mean, α_i the method effect and ε_{ij} an error or residual term, assumed to have a **normal distribution** with mean zero and variance σ^2. In terms of this model, interest lies in testing the null hypothesis that $\alpha_1 = \alpha_2 = \cdots = \alpha_k = 0$, where k is the number of groups. This is equivalent to testing the hypothesis that the k groups have the same mean.

Numerical example

In an experiment to compare different methods of teaching arithmetic, 36 students were divided randomly into 4 groups of equal size. At the end of the investigation, all students took a standard test, with the results

Method 1	Method 2	Method 3	Method 4
21	28	19	21
23	30	28	14
13	29	26	13
19	24	26	19
13	27	19	15
19	30	24	15
20	28	24	10
21	28	23	18
16	23	22	20

The analysis of variance table is

Source	SS	df	MS	F	P
Between methods	702.67	3	234.22	22.67	<0.001
Within methods	331.33	32	10.35		
Total	1034.00	35			

The results indicate a highly significant difference in the means of the four teaching methods. Identifying which particular methods differed would require more investigation using post hoc comparisons.

See also **analysis of covariance**, **parallel groups design** and **factorial designs**.

Analysis of variance table See **analysis of variance**.

ANCOVA Acronym for **analysis of covariance**.

ANOVA Acronym for **analysis of variance**.

A posteriori comparison Synonym for **post hoc comparison**.

Approximation A result that is not exact but is sufficiently close for required purposes to be of practical use. See also **age heaping** and **rounding**.

A priori comparisons Synonym for **planned comparisons**.

Arcsine transformation A transformation for a proportion *p*, designed to stabilize its variance and produce values more suitable for techniques such as analysis of variance.

Area under curve (AUC) Often a useful way of summarizing the information from a series of measurements made on an individual over time – for example, those collected in some longitudinal study. Essentially a weighted sum of individual measures, the weights being determined by the intervals between measurements. When the measurement intervals are regular, AUC is essentially equivalent to using the mean. See also **response feature analysis**.

Artificial intelligence A discipline that attempts to understand intelligent behaviour in the broadest sense, by getting computers to reproduce it, and to produce machines that behave intelligently, no matter what their underlying mechanism. (Intelligent behaviour is taken to include reasoning, thinking and learning.) See also **artificial neural network**. [*Representation and Processing of Spatial Expressions*, Mahwah, NJ, 1998. Lawrence Erlbaum].

Artificial neural network An applied statistical model designed in some simple respects to parallel the human neural network. Used by psychologists for attacking problems in areas such as pattern recognition, learning, language development and memory. Neural network models tend to be far more ambitious than traditional statistical models, and are often (but not always) more successful on large-scale problems. See also **artificial intelligence**. [*Substance Use and Misuse*, 1998, 33, 555–586].

Artificial pairing See **paired samples**.

Ascertainment bias A possible form of bias, particularly in a retrospective study, that arises from a relationship between the exposure to a risk factor and the probability of detecting an event of interest. In a study comparing women with cervical cancer and a control group, for example, an excess of oral contraceptive use amongst the cases might possibly be due to more frequent screening in this group. [*Neuropsychology*, 1995, 9, 209–210].

As randomized analysis Synonym for **intention-to-treat analysis**.

Assignment method Synonym for **discriminant analysis**.

Association A general term used to describe the relationship between two variables. Essentially synonymous with correlation. Most often applied in the context of binary variables forming a two-by-two contingency table. See also **phi-coefficient**.

Assumptions The conditions under which statistical techniques give valid results. For example, analysis of variance generally assumes normality, homogeneity of variance and independence of the observations.

Asymmetric distribution A probability distribution or frequency distribution which is not symmetrical about some central value. An example would be a distribution with positive skewness as shown in Figure 1, which displays the histogram of 200 reaction times (seconds) to a particular task.

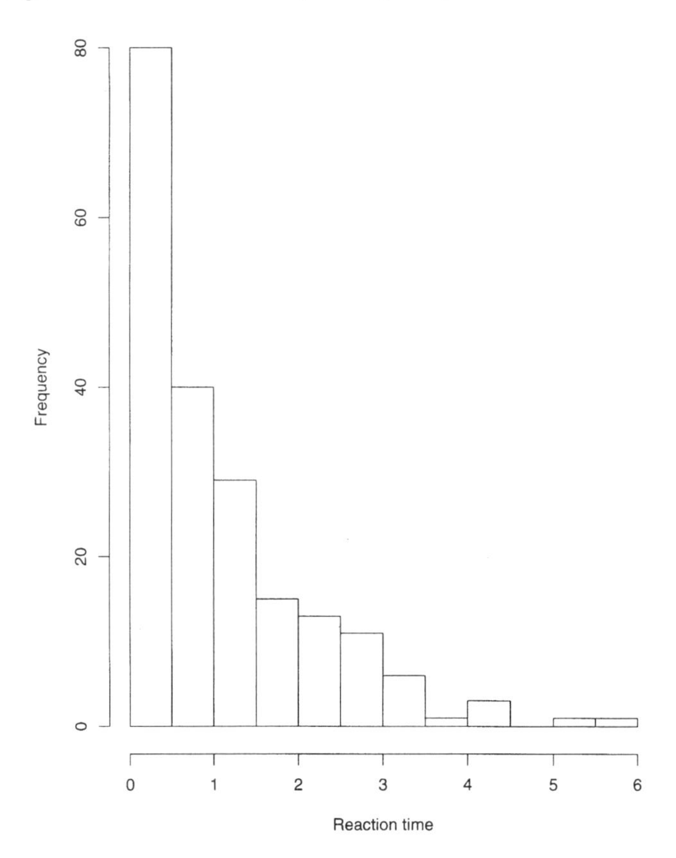

Figure 1 Frequency distribution of 200 reaction times.

Asymmetric proximity matrix A **proximity matrix** in which the off-diagonal elements, in the ith row and jth column and the jth row and ith column, are not necessarily equal. Such matrices are often the raw material of some forms of **multidimensional scaling**.

Mathematical details

An asymmetric proximity matrix for say N stimuli has the form

$$\mathbf{D} = \begin{bmatrix} \delta_{11} & \delta_{12} & \cdots & \delta_{1N} \\ \delta_{21} & \delta_{22} & \cdots & \delta_{2N} \\ \vdots & \vdots & \vdots & \vdots \\ \vdots & \vdots & \vdots & \vdots \\ \delta_{N1} & \delta_{N2} & \cdots & \delta_{NN} \end{bmatrix}$$

where $\delta_{ij} \neq \delta_{ji}$.

Numerical example

The following matrix has elements giving the number of citations of one journal by another for four well-known psychological journals.

Rows represent journals giving citation; columns represent citations received.

	1	2	3	4
1	31	10	10	1
2	7	235	55	0
3	16	54	969	28
4	3	2	30	310

1: *American Journal of Psychology*
2: *Journal of Abnormal Psychology*
3: *Journal of Abnormal and Social Psychology*
4: *Journal of Applied Psychology*

Attenuation A term applied to the estimated **correlation coefficient** between two variables when both are subject to measurement error, to indicate that the value of the correlation between the 'true values' is likely to be underestimated.

Attrition A term used to describe the loss of participants over the period of a **longitudinal study**. This may happen for a variety of reasons: for example, moving house and leaving the area, or becoming too unwell to attend. Such a phenomenon may cause problems in the analysis of data from such a study. See also **missing values**.

AUC Abbreviation for **area under curve**.

Autocorrelation The internal correlation of the observations in a **time series**, usually expressed as a function of the time lag between them.

Mathematical details

The autocorrelation at lag k, $\gamma(k)$, is defined mathematically as

$$\gamma(k) = \frac{E(X_t - \mu)(X_{t+k} - \mu)}{E(X_t - \mu)^2}$$

where X_t, $t = 0, \pm 1, \pm 2, \cdots$ represents the values of the series and μ is the mean of the series. E denotes **expected value**. The corresponding sample statistic is calculated as

$$\hat{\gamma}(k) = \frac{\sum_{t=1}^{n-k} (x_t - \bar{x})(x_{t+k} - \bar{x})}{\sum_{t=1}^{n} (x_t - \bar{x})^2}$$

where $\bar{x}$ is the mean of the observed series values $x_1, \ldots, x_n$. A plot of the sample values of the autocorrelation against the lag is known as the *autocorrelation function* or *correlogram* and is a basic tool in the analysis of time series, particularly for indicating possibly suitable models for the series.

Numerical example

The data below show the times recorded by the winners of the men's 1500 m races in the Olympic Games from 1948 to 1988:

Year	Winning time in seconds
1948	225.20
1952	225.20
1956	221.20
1960	215.60
1964	218.10
1968	214.90
1972	216.30
1976	219.20
1980	218.40
1984	212.50
1988	215.96

The data represent a short time series; they are plotted in Figure 2. The autocorrelations are given by:

Time interval between games (lag)	Correlation
4	0.48
8	0.01
12	−0.09

There is a substantial correlation between consecutive games separated by 4 years but only very small correlation for longer intervals. A plot of the autocorrelation (the ACF or correlogram) is shown in Figure 3.

[*Psychological Methods*, 1998, 3, 104–116].

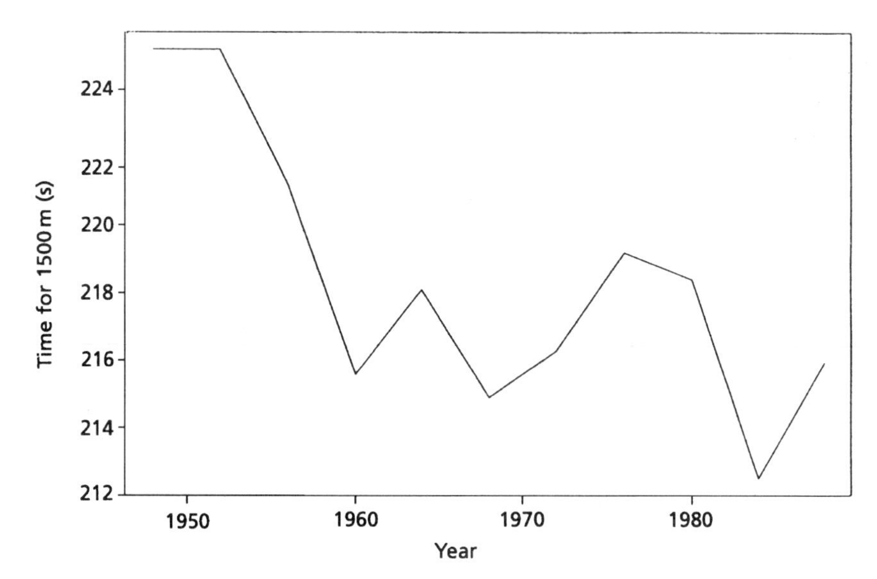

Figure 2 Plot of winning times in men's 1500 m races in the Olympic Games from 1948 to 1988.

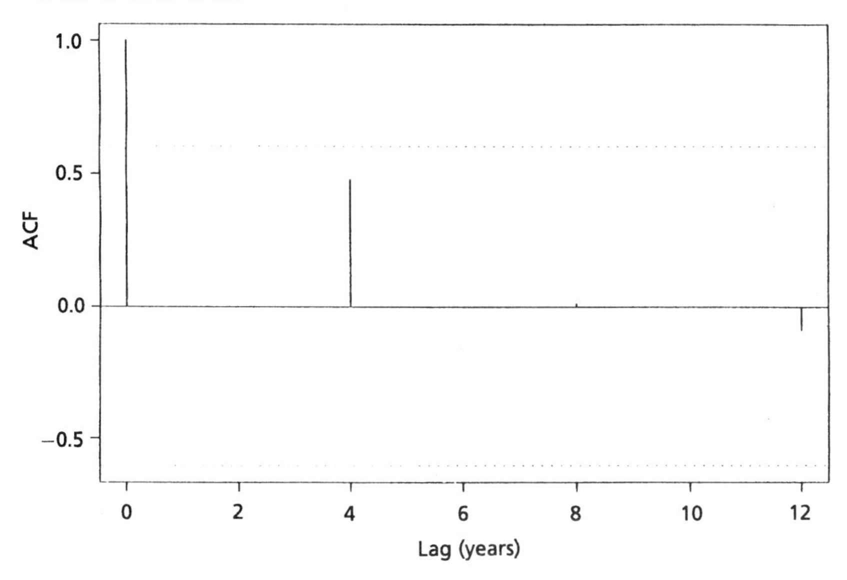

Figure 3 Autocorrelation function of winning times in men's 1500 m races in Olympic Games from 1948 to 1988.

Autocorrelation function See **autocorrelation**.

Automatic interaction detector (AID) A method that uses a set of categorical explanatory variables to divide data into groups that are relatively homogeneous with respect to the value of some dependent variable of interest. At each stage, the division of a group into two parts is defined by one of the explanatory variables, a subset of its categories defining one of the parts and the remaining categories the other part. Of the possible splits, the one chosen is that which maximizes the **between groups sum of squares** of the dependent variable. The groups eventually formed may often be useful in predicting the value of the dependent variable for some future observation. See also **classification and regression tree technique**. [*Mental Health in Black America.* Neighbors, Harold W (Ed); Jackson, James Sidney (Ed); et al. 1996 (pp 14–26). Thousand Oaks, CA: Sage].

Available case analysis An approach to handling **missing values** in a set of **multivariate data**, in which means, variances, covariances, etc. are calculated from all available subjects with non-missing values for the variable or pair of variables involved. Although this approach makes use of as much of the data as possible, it has disadvantages. One is that summary statistics will be based on different numbers of observations. More problematic, however, is that this method can lead to **variance—covariance matrices** and **correlation matrices** with properties that make them unsuitable for the application of many methods of **multivariate analysis** such as **principal components analysis** and **factor analysis**. See also **missing values**.

Average Most often used for the arithmetic mean of a sample of observations, but can also be used for other measures of location such as the median.

B

Backward elimination See **selection methods in regression**.

Balanced design A term usually applied to any experimental design in which the same number of observations is taken for each combination of the experimental factors. An **analysis of variance** of such a design is far simpler than when the separate cells of the design contain different numbers of observations (an *unbalanced design*) since unique sums of squares can be associated with each main effect and interaction. This is not so in an unbalanced design, for which different methods of calculating sums of squares may lead to different conclusions. See also **sequential sums of squares, Type I, Type II** and **Type III sums of squares**.

Balanced incomplete block design A design in which not all treatments are used in all **blocks**. Such designs have the following properties:

- Each block contains the same number of units.
- Each treatment occurs the same number of times in all blocks.
- Each pair of treatment combinations occurs together in a block the same number of times as any other pair of treatments.

In psychology, this type of design might be employed to avoid asking subjects to undertake an experimental procedure an unrealistic number of times, thus possibly preventing problems with **missing values**. For example, in a study with five experimental conditions (C_1, C_2, C_3, C_4 and C_5) it might be thought that subjects could realistically only be asked to make three visits. A possible balanced incomplete design in this case would be the following:

Subject	Visit 1	Visit 2	Visit 3
1	C_4	C_5	C_1
2	C_4	C_2	C_5
3	C_2	C_4	C_1
4	C_5	C_3	C_1
5	C_3	C_4	C_5
6	C_2	C_3	C_1
7	C_3	C_1	C_4
8	C_3	C_5	C_2
9	C_2	C_3	C_4
10	C_5	C_1	C_2

Bar chart A form of graphical representation for displaying data classified into a number of (usually unordered) categories. Equal-width rectangular bars are constructed over each category with height equal to the observed frequency of the category. Figure 4 gives an example involving counts of the number of occupants of 1469 cars (including the driver). See also **histogram** and **component bar chart**.

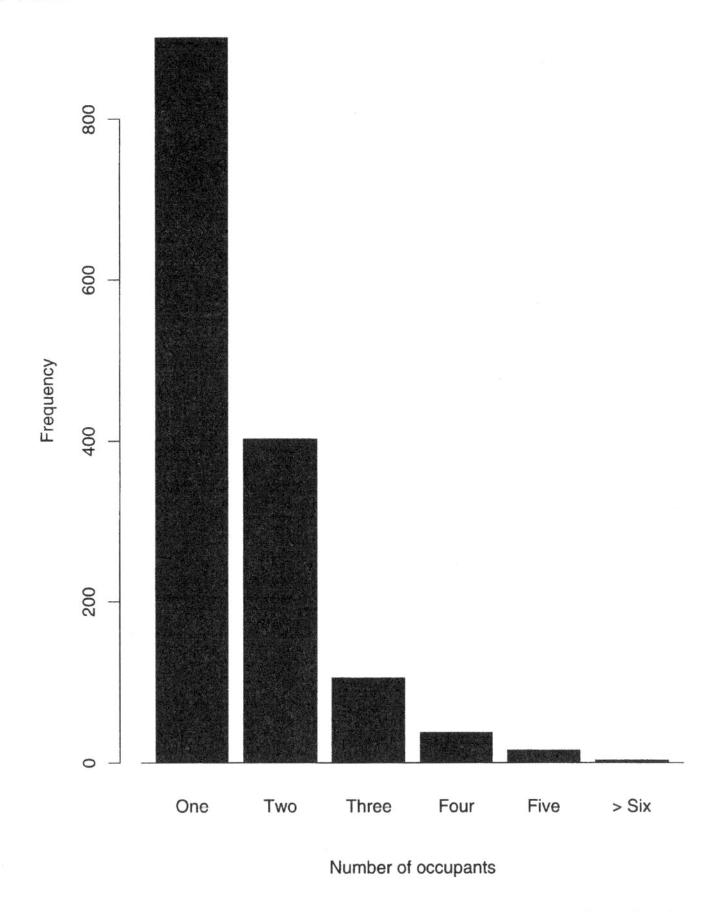

Figure 4 Bar chart of number of occupants (including the driver) of 1469 cars.

Bartlett's test A test for the equality of the variances of a number of populations. Sometimes used prior to applying **analysis of variance** to check the equality of variance assumption, although its usefulness is limited because of its known sensitivity to departures from normality against which analysis of variance is relatively robust. See also **Box's test**, **Hartley's test** and **Levine's test**.

Baseline characteristics Observations and measurements collected on participants at the time of entry into a study, before undergoing any experimental procedure. May include basic demographic information such as age, social class, sex etc., as well as recordings of the chosen dependent measure before the experimental procedure is applied. See also **change scores**.

Bayesian inference An approach to inference largely based on **Bayes' Theorem** and consisting of the following components:

(1) The **likelihood**, describing the process giving rise to the data in terms of a number of unknown parameters.
(2) The **prior distribution**, expressing what is known about the parameters prior to observing the data.

Bayes' theorem is used to combine these two components into the **posterior distribution** expressing what is known about the parameters after observing the data.

This form of inference differs from the classical form of **frequentist inference** in several respects, but in particular in the use of the prior distribution, which is absent from classical inference. This represents the investigator's knowledge about the parameters before seeing the data. Consequently, to a Bayesian, every problem is unique and is characterized by the investigator's belief about the parameters expressed in the prior distribution for the specific investigation. Classical statistics uses only the likelihood.

Bayes' Theorem A procedure for revising and updating the probability of some event in the light of new evidence. The theorem originates in an essay by the Reverend Thomas Bayes (1702–1761).

Mathematical details

In its simplest form the theorem may be written in terms of **conditional probabilities** as

$$\Pr(B|A) = \frac{\Pr(A|B)\Pr(B)}{\Pr(A)} = \frac{\Pr(A|B)\Pr(B)}{\Pr(A|B)\Pr(B) + \Pr(A|\bar{B})\Pr(\bar{B})}$$

where $\Pr(A|B)$ denotes the conditional probability of event A conditional on event B, and B and $\bar{B}$ are **complementary events**. The theorem gives the probability of B when A is known to have occurred.

Numerical example

A cab was involved in a hit and run accident at night. Two cab companies, the Green and the Blue, operate in the city. The following facts are known:

- 85% of the cabs in the city are Green and 15% are Blue.
- A witness identified the cab as Blue. The court tested the reliability of the witness under the same circumstances that existed on the night of the accident and concluded that the witness correctly identified each one of the two colours 80% of the time and failed 20% of the time.

What is the probability that the cab involved in the accident was actually Blue?

Kahneman and Tversky have found most people typically say the answer is around 80%; we can use Bayes' Theorem to find the correct answer. Equating the given percentages with probabilities and letting A be the event that a cab is Blue and B the event that the witness says he sees a Blue cab – then what we know is:

$$\Pr(A) = 0.15, \qquad \Pr(B|A) = 0.80$$

To apply Bayes' Theorem to get $\Pr(A|B)$, i.e. the probability that the cab is Blue given that the witness says it is blue, we need to calculate the unconditional probability that the witness says he sees a Blue cab, i.e. $\Pr(B)$. Since the witness is not infallible he will on occasions correctly identify a Blue cab and on others incorrectly identify a Green cab as Blue; consequently the required unconditional probability is given by the sum of the probabilities of these two events:

- Cab is Blue and witness correctly identifies cab as Blue – probability is 0.15×0.80.
- Cab is Green and witness incorrectly identifies cab as Blue – probability is 0.85×0.20. Therefore the required unconditional probability of the witness saying that he saw a Blue cab is

$$\Pr(B) = 0.15 \times 0.80 + 0.85 \times 0.20 = 0.29$$

Now we have all the terms necessary to apply Bayes' Theorem to give

$$\Pr(B|A) = \frac{0.80 \times 0.15}{0.29} = 0.41$$

So the probability that, given the witness identifying the cab as Blue, it actually *is* Blue is less than a half. Despite the eyewitness's evidence the hit-and-run cab is more likely to be Green than Blue! The evidence has, however, increased the probability that the offending cab is Blue from its value of 0.15, in the absence of any evidence, to 0.41.

Bernoulli distribution The **probability distribution** of a binary **random variable**, X, where

$$\Pr(X = 1) = p \quad \text{and} \quad \Pr(X = 0) = 1 - p$$

with Pr denoting probability. Named after Jacob Bernoulli (1654–1705).

Bernoulli trials A set of n independent **binary variables** in which the jth observation is either a 'success' or a 'failure', with the probability of success, p, being the same for all trials. Tossing a coin a particular number of times and associating 'success' with heads is an example. See also **binomial distribution.**

Beta coefficient A **regression coefficient** that is standardized so as to allow for a direct comparison between explanatory variables as to their relative explanatory power for a dependent variable. Calculated from the raw regression coefficients by multiplying them by the standard deviation of the corresponding explanatory variable and dividing by the standard deviation of the dependent variable.

Numerical example

Suppose that in a **multiple regression** involving two explanatory variables, x_1 and x_2, the regression coefficients are estimated to be -1.40 and 0.003. This might be taken as suggesting that x_1 is more important than x_2 in determining the dependent variable. But if the standard deviations of x_1 and x_2, 0.0083 and 16.42, and the standard deviation of the response variable, 0.066, are used to find the beta coefficients as:

$$x_1: \quad \text{beta coefficient} = \frac{-1.40 \times 0.0083}{0.066} = -0.18$$

$$x_2: \quad \text{beta coefficient} = \frac{0.03 \times 16.42}{0.066} = 0.76$$

then it is seen that it is actually variable x_2 which is of greater importance.

Beta (β) error Synonym for **Type II error**.

Between groups mean square See **analysis of variance**.

Between groups sum of squares See **analysis of variance**.

Between groups sums of squares and cross products matrix See **multivariate analysis of variance**.

Bias In general terms, deviation of results or inferences from the truth, or processes leading to such deviation. More specifically, the extent to which the statistical method used in a study does not estimate the quantity thought to be estimated, or does not test the hypothesis to be tested.

Biased estimator An estimator whose **expected value** is not the true value of the parameter.

An example

One possible estimator, of the population variance, σ^2, provided by a sample of observations $x_1, \ldots, x_n$, with mean $\bar{x}$, is s^2 given by

$$s^2 = \frac{1}{n} \sum_{i=1}^{n} (x_1 - \bar{x})^2$$

This has expected value $\left(\frac{n-1}{n}\right)\sigma^2$, *not* σ^2, and is therefore biased.

Bimodal distribution A **probability distribution**, or a **frequency distribution**, with two modes. Such distributions often indicate that the data consist of distinct groups of observations. Figure 5 shows an example of each.

Binary variable Observations which occur in one of two possible states, these often being labelled 0 and 1. Such data are frequently encountered in psychological investigations; commonly occurring examples include 'improved/not improved' and 'completed task/failed to complete task'. Data involving response variables of this type often require specialized techniques such as **logistic regression** for their analysis. See also **Bernoulli distribution**.

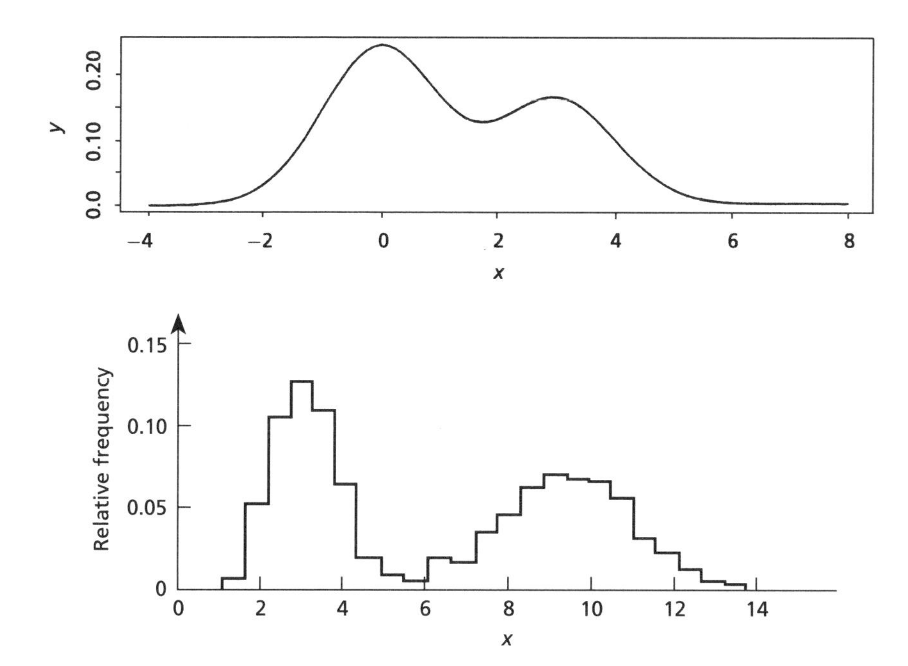

Figure 5 Bimodal probability (top) and frequency distribution (bottom).

Binomial distribution The **probability distribution** of the number of 'successes', X, in a series of independent trials, each of which can result in either a 'success' or a 'failure', when the probability of a success remains constant from trial to trial. A simple example is tossing a coin a number of times and equating 'success' with, for example, heads.

Mathematical details

$$\Pr(X = x) = \frac{n!}{x!\,(n-x)!}\,p^x(1-p)^{n-x}, \qquad x = 0, 1, 2, \ldots, n$$

where n is the number of trials and Pr denotes probability. The mean of such a variable is np and its variance is $np(1-p)$.

Numerical example

In tossing a fair coin 10 times the probability of obtaining 9 heads is

$$\Pr(X = 9) = \frac{10!}{9!\,1!}\left(\frac{1}{2}\right)^9\left(\frac{1}{2}\right)^1 = 0.00879$$

Bipolar factor A factor, resulting from the application of **factor analysis**, which has a mixture of positive and negative loadings. Such factors can be difficult to interpret, and attempts are often made to simplify them by the process of **factor rotation**. For an example see **factor analysis**.

Biserial correlation A measure of the strength of the relationship between two variables, one continuous (y) and the other recorded as a **binary variable** (x), but having underlying continuity and normality.

Mathematical details

The biserial correlation of a set of data is calculated as

$$r_b = \frac{\bar{y}_1 - \bar{y}_0}{s_y} \frac{pq}{u}$$

where $\bar{y}_1$ is the sample mean of the y variable for those individuals for whom $x = 1$, $\bar{y}_0$ is the sample mean of the y variable for those individuals for whom $x = 0$, s_y is the standard deviation of the y values, p is the proportion of individuals with $x = 1$, and $q = 1 - p$ is the proportion of individuals with $x = 0$. Finally u is the ordinate (height) of the **standard normal distribution** at the point of division between the p and q proportions of the curve.

Numerical example

The data below show the speech development level (0 = low, 1 = high) and IQ scores at age three of 20 infants. Here the categorical variable can be viewed as a surrogate for an underlying continuum of speech levels.

	IQ									
Low speech scores	87	90	94	94	97	103	103	104	106	108
High speech scores	100	103	106	112	113	114	114	118	119	120

The biserial correlation is given by $r_b = 0.978$, reflecting that, for these data, the high speech score group consistently have higher IQ scores than the other group.

See also **point biserial correlation**. [*British Journal of Social Psychology*, 1997, 36, 161–171].

Bit A unit of information, consisting of one binary digit.

Bivariate data Data consisting of measurements on each of two variables for a sample of individuals. For example, reaction time and anxiety level.

Bivariate distribution The **joint distribution** of two random variables, x and y. A well-known example is the *bivariate normal distribution*, which has the form

$$f(x, y) = \frac{1}{2\pi\sigma_1\sigma_2\sqrt{1-\rho^2}} \times \exp\left\{-\frac{1}{1-\rho^2}\left[\frac{(x-\mu_1)^2}{\sigma_1^2} - 2\rho\frac{(x-\mu_1)(y-\mu_2)}{\sigma_1\sigma_2} + \frac{(y-\mu_2)^2}{\sigma_2^2}\right]\right\}$$

where μ_1, μ_2, σ_1, σ_2 and ρ are, respectively, the means, standard deviations and **correlation coefficient** of the two variables. **Perspective plots** of such a distribution with zero means, variances equal to one, and various correlation values are shown in Figure 6.

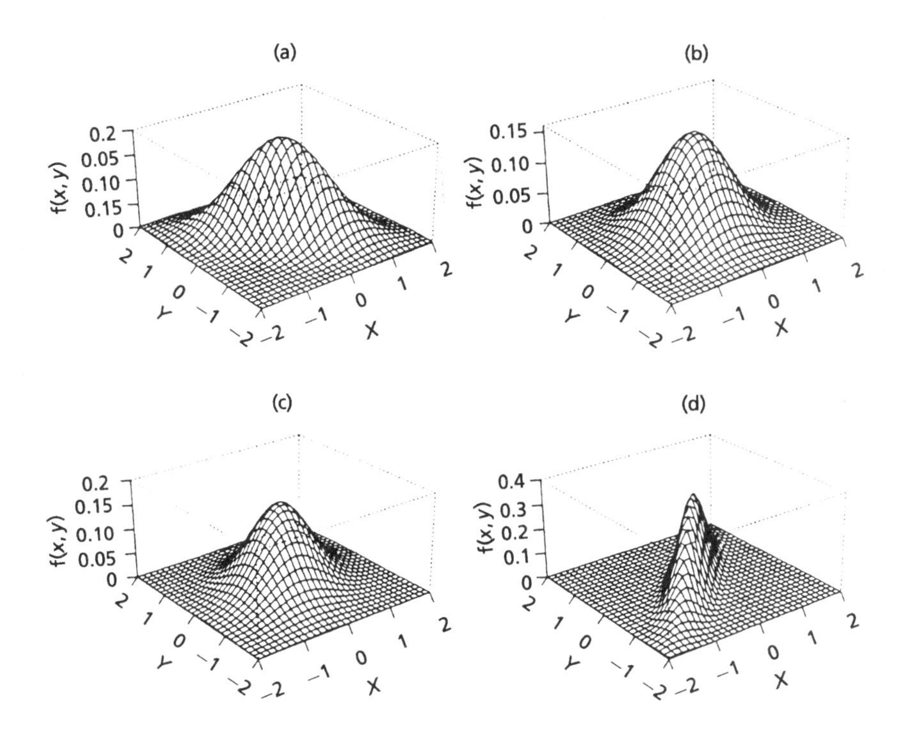

Figure 6 Perspective plots of four bivariate normal distributions with zero means, standard deviations equal to one, and correlations (a) −0.6, (b) 0.0, (c) 0.3, (d) 0.9.

Bivariate normal distribution See **bivariate distribution**.

Blinding A procedure commonly used in **clinical trials** in which the participant and/or the treating psychologist are kept unaware of whether the participant is receiving an active treatment or a placebo. If only the participant is unaware, the trial is said to be *single blind*; if both the participant and psychologist do not know the treatment, the trial is *double blind*. In trials when blinding to the psychologist is not possible, participant evaluations should be carried out by assessors who are blind to the treatment received. See also **bias**.

Block A term used in **experimental design** to refer to a homogeneous grouping of experimental units (often subjects) designed to enable the experimenter to isolate and, if necessary, eliminate variability due to extraneous causes. An example would be a subject in a **longitudinal study**.

Bonferroni correction A procedure for guarding against an increase in the probability of a **Type I error** when performing multiple significance tests. To maintain the probability of a Type I error at some selected value α, each of the m tests to be performed is judged against a significance level, α/m. For a small number of simultaneous tests (up to five) this method provides a simple and acceptable answer to the problem of multiple testing. It is, however, highly conservative and not recommended if large numbers of tests are to be applied, when one of the many other **multiple comparison procedures** available is generally preferable. See also **least significant difference test**, **Newman–Keuls test** and **Scheffé's**. [*Psychological Reports*, 1997, 80, 219–224].

Bootstrap A data-based simulation method for statistical inference, which can be used to study the variability of estimated characteristics of the **probability distribution** of a set of observations, and provide **confidence intervals** for parameters in situations where these are difficult or impossible to derive in the usual way. (The use of the term bootstrap derives from the phrase 'to pull oneself up by one's bootstraps'.) The basic idea of the procedure involves **sampling with replacement** to produce **random samples** of size n from the original data, $x_1, x_2, \ldots, x_n$; each of these is known as *a bootstrap sample* and each provides an estimate of the parameter of interest. Repeating the process a large number of times provides the required information on the variability of the estimator, and an approximate 95% confidence interval can, for example, be derived from the 2.5% and 97.5% **quantiles** of the replicate values.

Numerical example

The data below show the times in seconds taken by children in two experimental groups to construct a pattern from nine coloured blocks taken from the Wechsler Intelligence Scale for Children (WISC). The two groups were given different instructions for the task. The 'row group' were told to start with a row of three blocks, and the 'corner group' were told to start with a corner of three blocks.

Row group: 675, 510, 490, 850, 317, 464, 525, 298, 491, 196, 268, 372, 370, 739, 430, 410
Corner group: 342, 222, 219, 513, 295, 285, 408, 543, 298, 494, 317, 407, 290, 301, 325, 360

Construction of confidence intervals for the WISC data using bootstrap proceeds as follows:

- The procedure is based on drawing random samples of 16 observations with replacement from each of the row and corner groups.
- The random samples are found by labelling the observations in each group with integers $1, 2, \ldots, 16$ and selecting random samples of these integers (with replacement).
- The mean difference or median difference is calculated for each bootstrap sample.
- The histograms for mean differences and median differences of 1000 bootstrap samples are shown in Figure 7.

An approximate 95% confidence interval can be derived from the 25th and 975th largest of the 1000 replicates. This gives:
– mean: (16.06, 207.81)
– median: (7.5, 193.00).

See also **jackknife**. [*Educational and Psychological Measurement*, 1998, 58, 221–240].

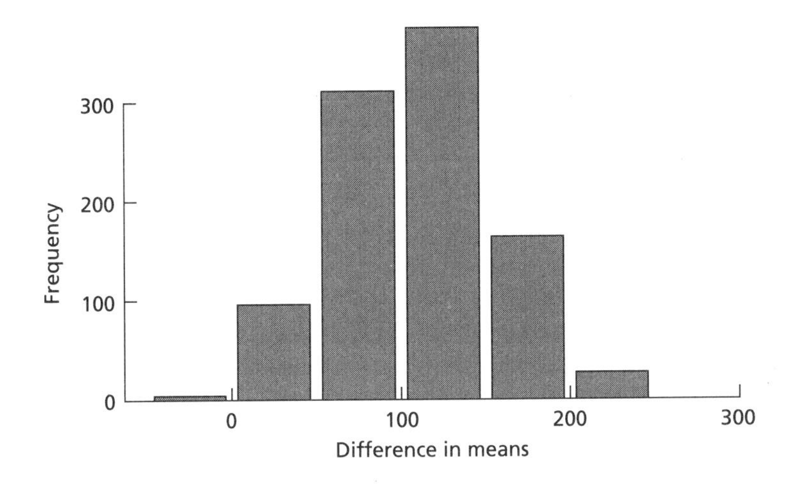

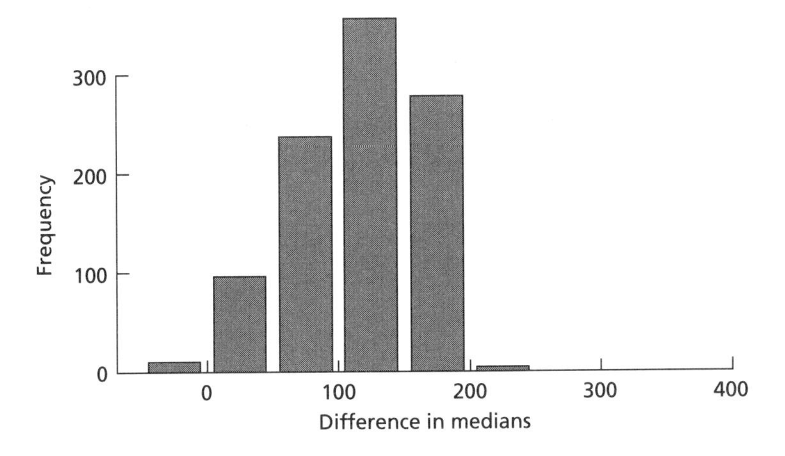

Figure 7 Histograms of mean and median differences found from applying the bootstrap.

Bootstrap sample See **bootstrap**.

Bowker's test A test that can be applied to **square contingency tables** to assess the hypothesis that the probability of being in cell j,i is equal to the probability of being in cell i,j.

Mathematical details

Under the hypothesis stated above, the **expected frequencies** in both cell i,j (E_{ij}) and cell j,i (E_{ji}) are given by:

$$E_{ij} = E_{ji} = \frac{n_{ij} + n_{ji}}{2}$$

where n_{ij} and n_{ji} are the corresponding observed frequencies. The **test statistic** is

$$\chi^2 = \sum_{i<j} \frac{(n_{ij} - n_{ji})^2}{n_{ij} + n_{ji}}$$

Under the hypothesis of symmetry, χ^2 has approximately a **chi-squared distribution** with $c(c-1)/2$ degrees of freedom, where c is the number of rows of the table (and the number of columns). In the case of a **two-by-two contingency table**, the procedure is equivalent to **McNemar's test**.

Numerical example

The data below were collected in a study of social mobility in the United Kingdom:

Social class of fathers and their sons

	Sons' class		
Fathers' class	**Upper**	**Middle**	**Lower**
Upper	588	395	159
Middle	349	714	447
Lower	114	320	411

To test the hypothesis that the changes in class between fathers and sons occur in both directions with the same probability, Bowker's test can be applied:

$$\chi^2 = \frac{(349-395)^2}{349+395} + \frac{(159-114)^2}{159+114} + \frac{(447-320)^2}{447+320} = 31.29$$

This has three degrees of freedom and is highly significant. The hypothesis of symmetry is rejected. Here this reflects the observed downward drift in class.

Box-and-whisker plot A graphical method of displaying the important characteristics of a set of observations. The display is based on the **five-number summary** of the data, with the 'box' part covering the **inter-quartile range**, and the 'whiskers' extending to include all but **outside observations**, these being indicated separately. Such graphics are particularly useful when comparing different groups of observations. An example involving the IQ scores of a number of psychologists and a number of statisticians is shown in Figure 8.

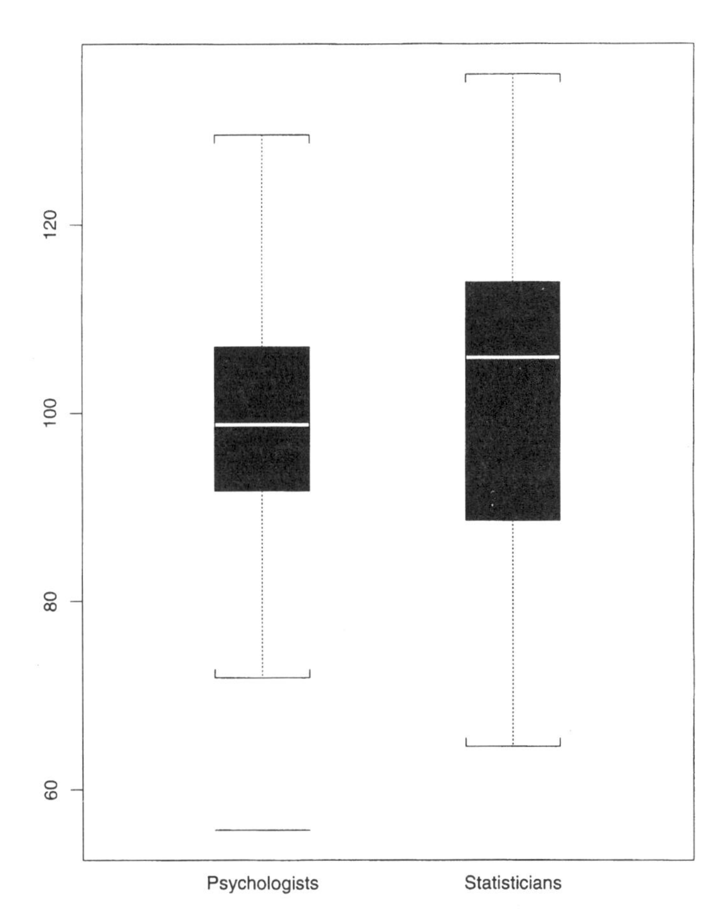

Figure 8 Box-and-whisker plots of IQ scores of a sample of psychologists and a sample of statisticians.

Box plot Synonym for **box-and-whisker plot**.

Box's test A test for assessing the equality of the variances in a number of populations that is less sensitive to departures from normality than **Bartlett's test**. See also **Hartley's test** and **Levine's test**.

Bubble plot A method for displaying observations which involve three variable values. Two of the variables are used to form a **scatter diagram**, and values of the third variable are represented by circles with differing radii centred at the appropriate position. An example of this type of plot is given in Figure 9. The scatter diagram is formed from 'price' and 'temperature' over a 30-week period, with the size of the circles representing consumption of ice cream.

Bump hunting A colourful term for the examination of **frequency distributions** for local maxima or modes that might be indicative of separate groups of subjects. See also **finite mixture distribution** and **bimodal distribution**.

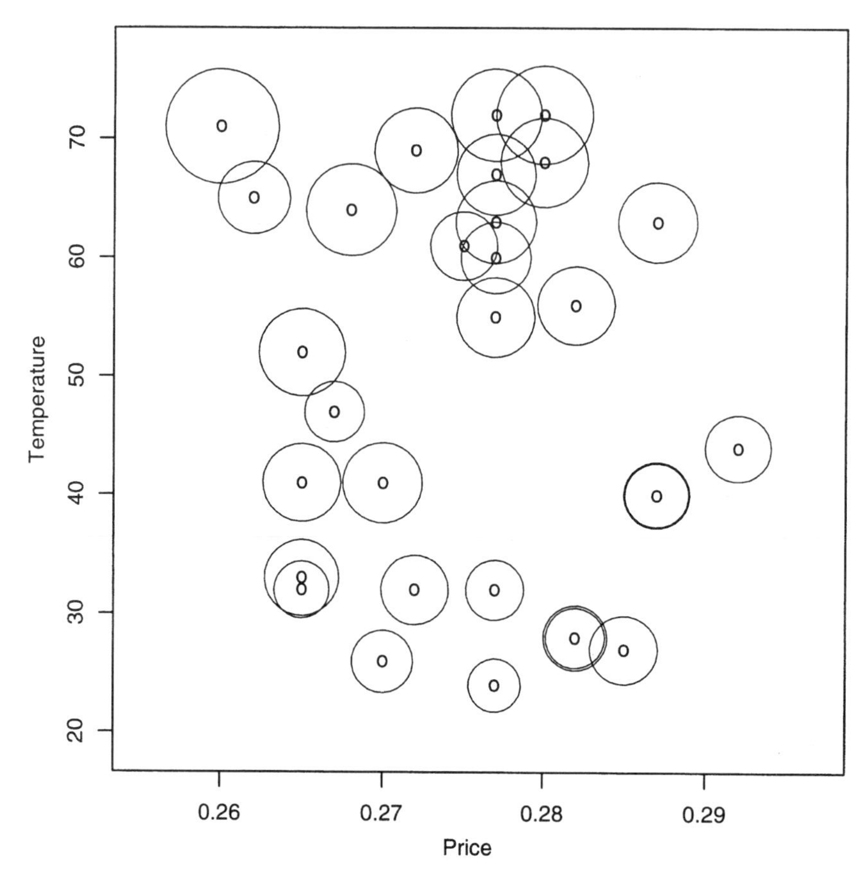

Figure 9 Bubble plot of price, temperature and consumption of ice cream in 30 one-week periods.

Byte A unit of information, as used in digital computers, equal to eight **bits**.

C

Caliper matching See **matching**.

Canonical correlation analysis A method of analysis for exploring the relationships between two distinct sets of variables. It might, for example, be of interest to investigate whether there is a relationship between achievement in arithmetic, reading and spelling as measured in elementary school and a set of variables reflecting early childhood development, such as age of first speech, walking and toilet training. In many respects the method can be viewed as an extension of **multiple regression** to situations involving more than a single dependent variable. Alternatively it can be considered as analogous to **principal components analysis**, except that a correlation rather than a variance is maximized.

Mathematical details

The aim is to find the linear functions of one set of variables $x_1, x_2, \ldots, x_q$ that maximally correlate with linear functions of another set of variables $y_1, y_2, \ldots, y_r$. Deriving the coefficients to define these linear functions involves, essentially, finding the **eigenvalues** and **eigenvectors** of the matrix $\mathbf{R}$ given by

$$\mathbf{R} = \mathbf{R}_{yy}^{-1}\,\mathbf{R}_{yx}\,\mathbf{R}_{xx}^{-1}\,\mathbf{R}_{xy}$$

where $\mathbf{R}_{yy}$ is the **correlation matrix** of the y variables, $\mathbf{R}_{xx}$ is the correlation matrix of the x variables, and $\mathbf{R}_{xy} = \mathbf{R}_{yx}$ contains the correlations between the x and y variables.

Numerical example

The data below show the correlations between grade point averages and SAT scores (variables 3–6) and university grade point averages (variables 1 and 2), for 406 students. Interest lies in assessing the relationship between the university grades and the other variables.

University Year 1	1.000					
University Year 2	0.812	1.000				
Grade 10	0.191	0.189	1.000			
Grade 11	0.221	0.199	0.750	1.000		
Grade 12	0.421	0.390	0.390	0.410	1.000	
SAT Score	0.230	0.180	0.020	0.030	0.180	1.000

The derived canonical linear functions are

Variable	CV1	CV2
University Year 1	0.783	−1.522
University Year 2	0.254	1.693
Variable	**CV1**	**CV2**
Grade 10	−0.002	0.898
Grade 11	0.146	−0.797
Grade 12	0.814	0.395
SAT Score	0.342	−0.745
Correlations	0.46	0.07

The conclusion is that a suitably standardized average of high school grades is a reasonable predictor of the average grades of the first two years at university. Change in high school marks does not appear to be correlated with change in university marks.

[*Journal of Applied Psychology*, 1998, 83(3), 462–470].

Carryover effects See **crossover design**.

CART Abbreviation for **classification and regression tree technique**. [*Journal of Abnormal Psychology*, 1997, 106, 586–597].

Cartogram A diagram in which descriptive statistical information is displayed on a geographical map by means of shading or by using a variety of different symbols. Figure 10 shows life expectancy dichotomized to <70 years and ≥ 70 years, for the states in the USA.

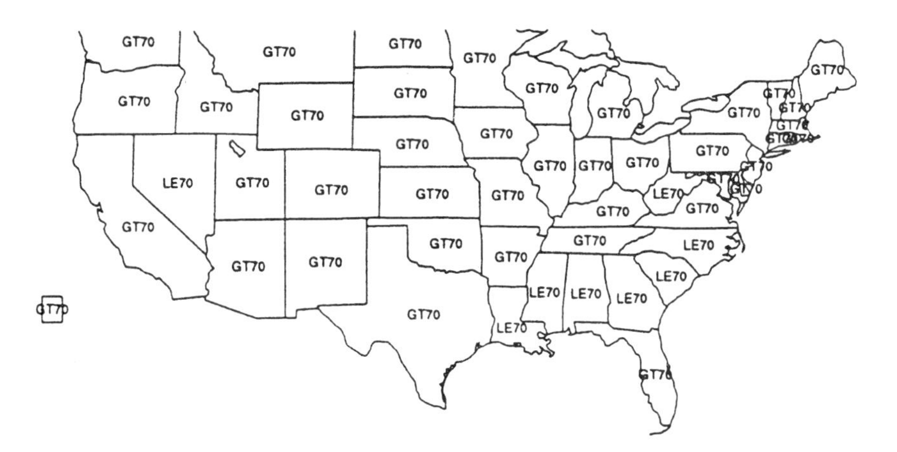

Figure 10 Cartogram of life expectancy in the USA.

Case–control study See **retrospective study**.

Categorical variable A variable that gives the appropriate label of an observation after allocation to one of several possible categories: for example, gender (male or female), marital status (married, single or divorced). The categories are often given numerical labels, but for this type of data these have no numerical significance. See also **binary variable**, **continuous variable** and **ordinal variable**.

Causality The relating of causes to the effects they produce. Many investigations in psychology seek to establish causal links between events: for example, early childhood separation and depression in adulthood. In general, the strongest claims to have established causality come from data collected in **experimental studies**. Relationships established in **observational studies** may be very suggestive of a causal link, but are always open to alternative explanations.

Ceiling effect A term used to describe what happens when many participants in a study have scores on a variable that are at or near the possible upper limit ('ceiling'). Such an effect may cause problems for some types of analysis, because it reduces the possible amount of variation in the variable. The converse, or *floor effect*, causes similar problems. A psychologist might, for example, administer a memory test which was so easy for one group that all the scores lay between 90% and 100%, or alternatively it might be so difficult that no-one scored more than 5%. [*Psychology, Public Policy and Law*, 1997, 3, 381–401].

Censored observations An observation x_i is said to be censored if it is known that, given L_i and U_i, the exact value of x_i is unknown, only that $x_i \leq U_i$ or $x_i \geq L_i$. Such observations arise most frequently in studies where the main dependent variable is time until a particular event occurs (for example, time to death) when, at the completion of the study, the event of interest has not happened to a number of subjects. In psychology, number of trials to learn a particular task, when a maximum number of trials is allowed, during which some subjects do not learn the task, is an example.

Census A study that involves the observation of every member of a population.

Central limit theorem If a **random variable** y has population mean μ and population variance σ^2, then the sample mean, $\bar{y}$, based on n observations, has an approximate **normal distribution** with mean μ and variance σ^2/n, for sufficiently large n. The examples shown in Figure 11 are based on samples of size 2, 5, 10 and 50 from a **uniform distribution** in the interval (0, 1). As sample size increases, the distribution of the arithmetic mean gets closer to the form of a normal distribution.

Central range The range within which the central 90% of values of a set of observations lie.

Central tendency A property of the **frequency distribution** of a variable, usually measured by statistics such as the mean, median and mode.

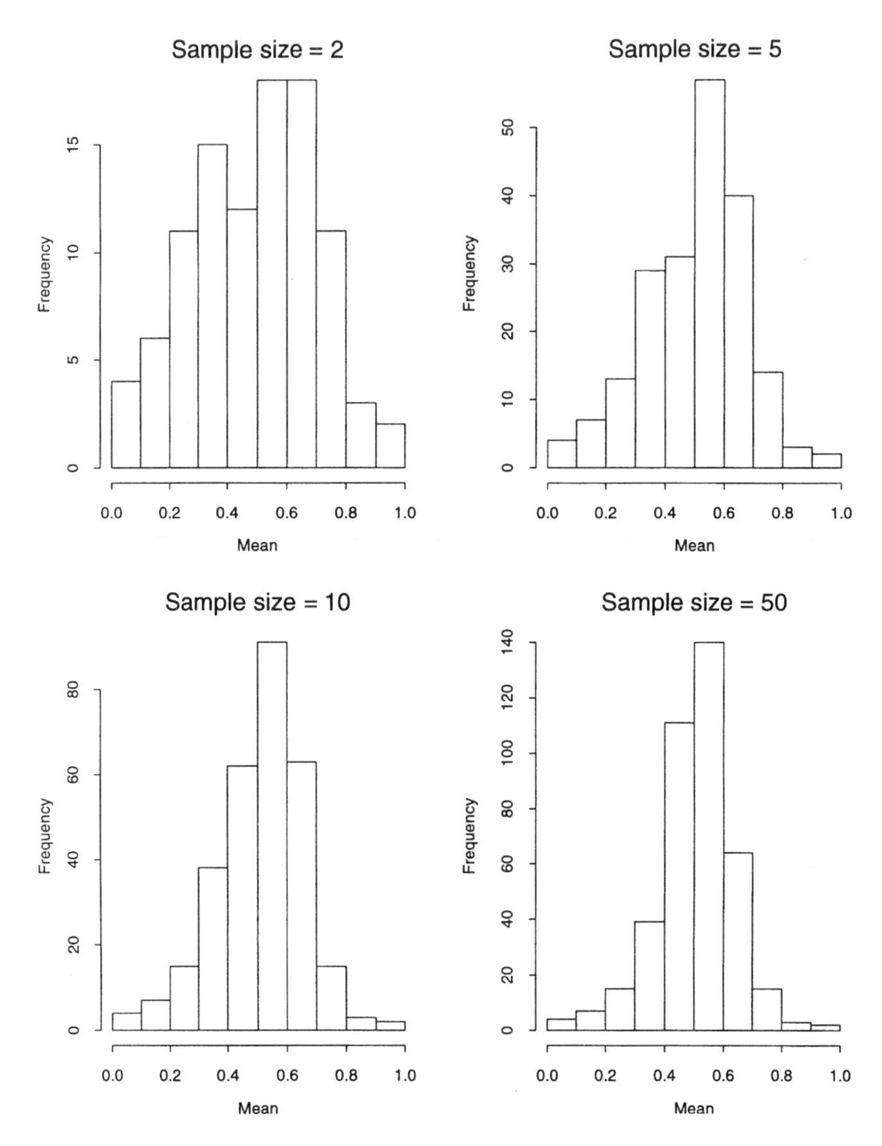

Figure 11 Examples of the operation of the central limit theorem. Samples taken from a uniform distribution in the interval (0,1).

CFA Abbreviation for **confirmatory factor analysis**.

Chaining A phenomenon often encountered in the application of single linkage clustering, which relates to the tendency of the method to incorporate intermediate points between distinct clusters into an existing cluster rather than initiate a new one. Figure 12 illustrates the problem.

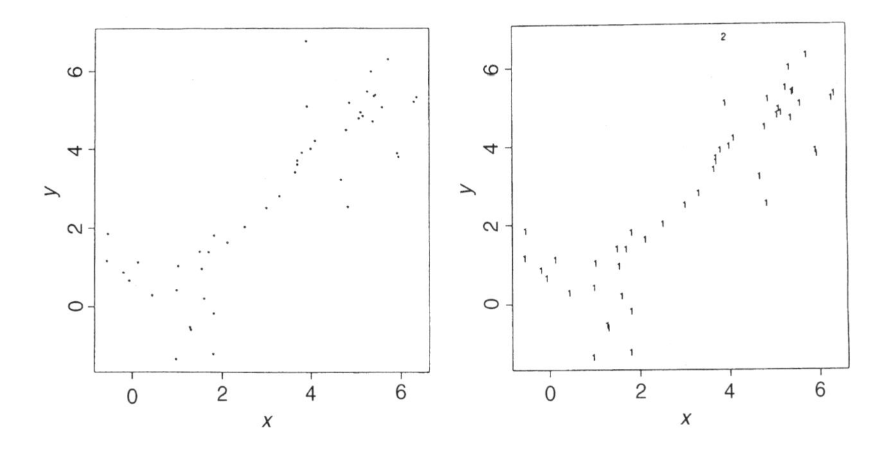

Figure 12 An illustration of chaining. The illustration on the right shows the results of single linkage clustering for two groups.

Change scores Scores obtained by subtracting a post-treatment score on some dependent variable from the corresponding pre-treatment, baseline value. Commonly used in the analysis of 'pre–post' designs although well known to be less powerful than analysis of covariance applied to the post score with the pre value as covariate. See also **adjusting for baseline** and **baseline balance**.

Chernoff's faces A technique for representing multivariate data graphically. Each observation is represented by a computer-generated face, the features of which are controlled by an observation's variable values. The collection of faces representing the set of observations may be useful in identifying groups of similar individuals, outliers, etc. Figure 13 shows an example. See also **glyphs**.

Chi-squared distribution The probability distribution of the sum of squares of a number (v) of independent variables each having a standard normal distribution. Used in many areas of statistics, especially when testing a particular hypothesis about categorical variables: for example, independence in a two-dimensional contingency table.

Mathematical details

The mathematical formula for the chi-squared distribution is

$$f(x) = \frac{1}{2^{v/2}\Gamma(v/2)} x^{(v-2)/2} e^{-x/2}, \qquad x > 0$$

where Γ is the gamma function. This is essentially a gamma distribution with $x = v/2$.

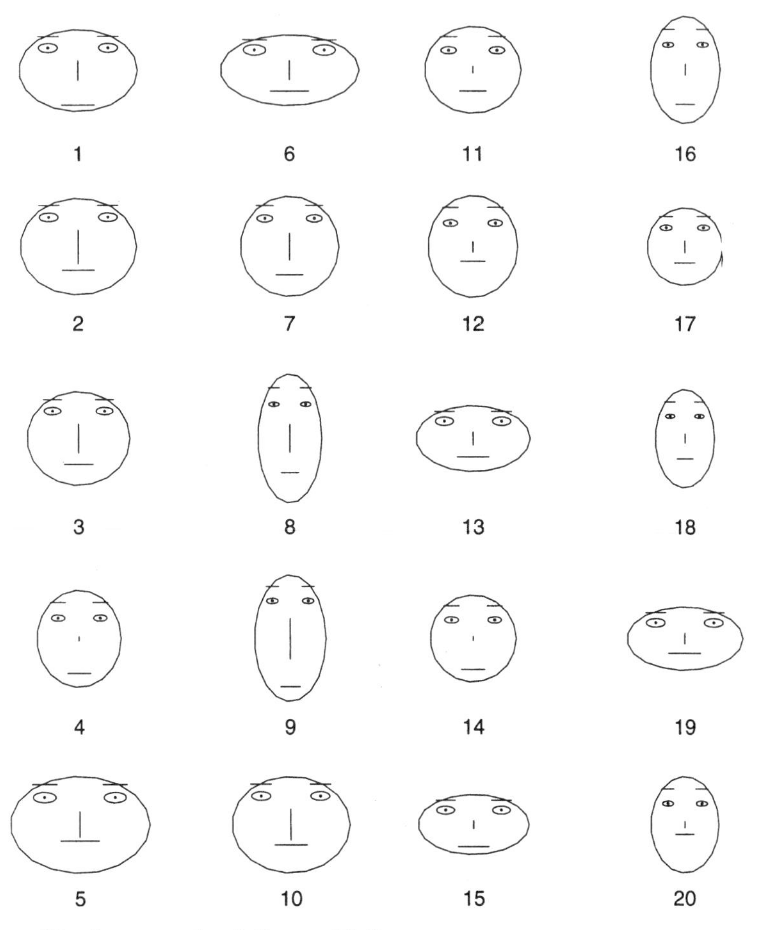

Figure 13 An example of Chernoff's faces.

Chi-squared statistic A statistic having, at least approximately, a **chi-squared distribution**.

Mathematical details

An example is the **test statistic** used to assess the independence of the two variables forming a **contingency table** with r rows and c columns:

$$\chi^2 = \sum_{i=1}^{r} \sum_{j=1}^{c} \frac{(O_{ij} - E_{ij})^2}{E_{ij}}$$

where O_{ij} represents the observed value in the ijth cell and E_{ij} the **expected frequency** under independence. Under the hypothesis of independence, χ^2 has, approximately, a chi-squared distribution with $(r-1)(c-1)$ degrees of freedom. For a numerical example see **contingency tables**.

χ^2 statistic Synonymous with **chi-squared statistic**.

Chi-squared test for trend A test applied to a **two-dimensional contingency table** in which one variable has two categories and the other has k ordered categories, to assess whether there is a difference in the trend of the proportions in the two groups. The result of using the ordering in this way is a test that is more powerful than using the **chi-squared statistic** to test for independence.

Mathematical details

The test relies on assigning scores to both the row and column categories and then calculating regression coefficients in the usual way. If b_{yx} is the estimated regression coefficient and $V(b_{yx})$ its estimated variance then the component of chi-square due to linear trend is given by

$$\frac{b_{yx}^2}{V(b_{yx})}$$

Numerical example

The data below consist of a sample of 223 boys classified according to age and to whether or not they had been rated as inveterate liars:

	Age group					
	5–7	**8–9**	**10–11**	**12–13**	**14–15**	
Inveterate liars	6	18	19	27	25	95
Non-liars	15	31	31	32	19	128
Total	21	49	50	59	44	223

Here, an examination of the proportion of inveterate liars in each age group, namely

0.286 0.367 0.380 0.458 0.568

suggests that the proportion increases steadily with age. Allocating scores of −2, −1, 0, 1, 2 to the age groups and 0 to non-liars and 1 to inveterate liars, the regression coefficient is estimated to be 0.065 with variance 0.00069. The overall chi-square of 6.69 with 4 degrees of freedom (P-value 0.15) can be partitioned as follows:

	df	χ^2
Linear regression of lying on age	1	6.19
Departure from regression	3	0.50
		6.69

The regression component has an associated P-value of 0.02, indicating a significant association between lying and age – here an increase in lying with age.

CI Abbreviation for **confidence interval**.

Class frequency The number of observations in a class interval of the observed **frequency distribution** of a variable.

Class intervals The intervals of the **frequency distribution** of a set of observations.

Classical inference Synonym for **frequentist inference**.

Classification and regression tree technique (CART) An alternative to **multiple regression** and associated techniques, for determining subsets of explanatory variables most important for the prediction of the response variable. Rather than fitting a model to the sample data, a 'tree structure' is generated by dividing the sample recursively into a number of groups, each division being chosen so as to maximize some measure of the difference in the response variable in the resulting two groups. The resulting structure often provides easier interpretation than a regression equation, as those variables most important for prediction can be quickly identified. Additionally, this approach does not require distributional assumptions and is also more resistant to the effects of **outliers**.

Mathematical details

At each state the sample is split on the basis of a single variable, x_i, or a set of variables $x_1, x_2, \ldots, x_q$, according to the answers to such questions as 'Is $x_i \leq c$' (univariate split), 'is $\sum_{i=1}^{q} a_i x_i \leq c$' (**linear function** split) and 'does $x_i \in A$' (if x_i is a categorical variable).

[*Journal of Abnormal Psychology*, 1997, 106, 586–597].

Classification matrix A term often used in **discriminant analysis** for the matrix containing counts of correct classifications on the main diagonal and incorrect classifications elsewhere.

Classification rule See **discriminant analysis**.

Classification techniques A generic term used for both **cluster analysis** methods and **discriminant analysis**, although more widely applied to the former.

Clinical change indicator Synonym for **reliable change indicator**.

Clinical trial A **prospective study** involving human participants, designed to determine, for example, the effectiveness of a treatment, a teaching technique, or a therapeutic regimen administered to people with specific disorders: for example, phobias. The gold standard is the double blind randomized controlled trial in which people are randomly allocated to treatments and neither they nor the treating psychologist know which treatment they receive. [*Journal of Consulting and Clinical Psychology*, 1998, 66, 429–433].

Cluster analysis A set of methods for constructing a (hopefully) sensible and informative classification of an initially unclassified set of data, using the variable values observed on each individual. See also **agglomerative hierarchial clustering**, **K-means cluster analysis**, **finite mixture distribution**, and **single linkage clustering**. [*Journal of Abnormal Psychology*, 1998, 107, 3, 412–422].

Clustered data A term applied to both data in which the sampling units are grouped together into clusters showing some common feature – for example, animal litters, families or geographical regions – and **longitudinal data** in which a cluster relates to the set of repeated measurements on the same individual. A distinguishing feature of such data is that they tend to exhibit intra-cluster correlation, and their analysis needs to address this correlation to reach valid conclusions. Methods of analysis that ignore the correlations tend to be inadequate. In particular they are likely to give standard errors that are too low. [*American Journal of Community Psychology*, 1994, 22, 595–615].

Cluster randomization The **random allocation** of groups or clusters of individuals in the formation of treatment groups. For example, in comparing a number of teaching methods, entire classes may be randomized to receive a particular method. Although not as statistically efficient as individual randomization, the procedure frequently offers important economic, feasibility or ethical advantages. [*American Journal of Public Health*, 1995, 85, 1378–1383].

Cluster sampling A method of sampling in which the members of a population are arranged in groups (the 'clusters'). A number of clusters are selected at random and those chosen are then subsampled. The clusters generally consist of natural groupings: for example, families, hospitals, schools, etc. See also **random sample**.

Coarse data A term sometimes used when data are neither entirely missing nor perfectly present. A common situation where this occurs is when the data are subject to **rounding**; others correspond to **digit preference** and **age heaping**.

Cochrane collaboration An international network of individuals committed to preparing, maintaining and disseminating systematic reviews of the effects of health care. The collaboration is guided by six principles: collaboration, building on people's existing enthusiasm and interests, minimizing unnecessary duplication, avoiding bias, keeping evidence up to date and ensuring access to evidence. Most concerned with the evidence from randomized **clinical trials**. See also **evidence-based medicine**.

Cochran's *Q*-test A procedure for assessing the hypothesis of no inter-observer **bias** in some situations where a number of raters judge the presence or absence of some characteristic on a number of subjects. Essentially a generalized **McNemar's test**.

Mathematical details

The **test statistic** is given by:

$$Q = \frac{r(r-1)\sum_{j=1}^{r}(y_{.j} - y_{..}/r)^2}{r y_{..} - \sum_{i=1}^{n} y_{i.}^2}$$

where $y_{ij} = 1$ if the ith person is judged by the jth rater to have the characteristic present and 0 otherwise, $y_{i.}$ is the total number of raters who judge the ith participant to have the characteristic, $y_{.j}$ is the total number of participants the jth rater judges as having the characteristic present, $y_{..}$ is the total number of 'present' judgements made, n is the number of participants and r is the number of raters. If the hypothesis of no inter-observer bias is true, Q has, approximately, a **chi-squared distribution** with $r - 1$ degrees of freedom.

Numerical example

The data below were collected in a study comparing the symptomatology of eight schizophrenic patients as judged by five psychologists in an interview. The symptom involved in these data is religious preoccupation, rated 0 if thought to be absent and 1 if considered present.

	Psychologist				
Patient	**1**	**2**	**3**	**4**	**5**
1	0	0	0	0	0
2	0	0	0	0	1
3	0	0	0	0	0
4	0	0	0	0	0
5	0	0	1	0	0
6	0	0	1	1	1
7	0	0	0	0	0
8	1	0	1	1	1

For these data

$$Q = 5 \times 4 \times \frac{(1 - 1.8)^2 + (0 - 1.8)^2 + (3 - 1.8)^2 + (2 - 1.8)^2 + (3 - 1.8)^2}{45.0 - 27.0}$$

$$= 7.55$$

If the hypothesis of no inter-observer bias is true, Q has a chi-squared distribution with 4 degrees of freedom. The associated **P-value** is 0.11, so the hypothesis is accepted, indicating a lack of **bias** in the ratings of the five psychologists.

[*Psychomusicology*, 1995, 14, 35–52].

Coefficient of alienation A name sometimes used for $1 - r^2$, where r is the estimated value of the **correlation coefficient** of two random variables. See also **coefficient of determination**.

Coefficient of concordance A coefficient used to assess the agreement among m raters ranking n individuals according to some specific characteristic.

Mathematical details

The coefficient, W, is defined as

$$W = \frac{12S}{m^2(n^3 - n)}$$

where S is the sum of squares of the differences between the total of the ranks assigned to each individual and the value $m(n+1)/2$. W can vary from 0 to 1, with the value 1 indicating perfect agreement.

Numerical example

Here the data give the relative ranks assigned by four psychologists to eight mothers based on the child-centred nature of their language.

	Psychologist				
Mother	**1**	**2**	**3**	**4**	**Total of ranks**
1	1	2	1	3	7
2	3	1	2	2	8
3	4	4	3	1	12
4	5	3	5	4	17
5	2	5	4	5	16
6	8	7	6	8	29
7	6	8	7	7	28
8	7	6	8	6	27

In this case $m = 4$ and $n = 8$.

$$S = (7-18)^2 + (8-18)^2 + (12-18)^2 + (17-18)^2 + (16-18)^2 + (29-18)^2 + (28-18)^2 + (27-18)^2 = 564$$

so that

$$W = 12 \times 564/16(512 - 8) = 0.839$$

indicating a reasonably high level of agreement.

[*British Journal of Clinical Psychology*, 1994, 33, 208–210].

Coefficient of determination The square of the **correlation coefficient** between two variables x and y. Gives the proportion of the variation in one variable that is accounted for by the other. For example a correlation of 0.8 implies that 64% of the variance of y is accounted for by x.

Coefficient of variation A measure of spread for a set of data, defined as

$$100 \times \text{standard deviation/mean}$$

Originally proposed as a way of comparing the variability in different distributions, but found to be sensitive to errors in the mean.

Cohort See **cohort study**.

Cohort study An investigation in which a group of individuals (the *cohort*) is identified and followed prospectively, perhaps for many years, and their subsequent behaviour of interest recorded. The cohort may be subdivided at the onset into groups with different characteristics, for example, exposed and not exposed to some risk factor, and at some later stage a comparison made of the incidence of a particular disease in each group. A recent example involves studying the cognitive functioning of farmers regularly exposed to sheep dip chemicals. See also **longitudinal study** and **prospective study**. [*Social Psychiatry and Psychiatric Epidemiology*, 1998, 33, 57–65].

Coincidences Surprising occurrence of events, perceived as meaningfully related, with no apparent causal connection. Such events abound in everyday life and are often the source of some amazement. As pointed out by Fisher, however, 'the one chance in a million will undoubtedly occur, with no less and no more than its appropriate frequency, however surprised we may be that it should occur to us'.

Collapsing categories A procedure often applied to contingency tables in which two or more row or column categories are combined, in many cases so as to yield a reduced table in which there are a larger number of observations in particular cells. Not to be recommended in general since it can result in misleading conclusions. See also **Simpson's paradox**.

Collinearity See **multicollinearity**.

Common factors See **factor analysis**.

Common factor variance A term used in factor analysis for that part of the variance of a variable shared with the other observed variables via the relationships of these variables to the common factors. Often known as communality.

Community controls See **control group**.

Community intervention study An intervention study in which the experimental unit to be randomized to different treatments is not an individual but a group of people: for example, a school or a factory. See also **cluster randomization**.

Comparison group Synonym for **control group**.

Comparisonwise error rate Synonym for **per-comparison error rate**.

Complementary events Mutually exclusive events A and B for which:

$$\Pr(A) + \Pr(B) = 1$$

where Pr denotes probability. For example, Pr(male) + Pr(female) = 1.

Complete case analysis An analysis that uses only individuals who have a complete set of measurements. Any individual with one or more missing values is not included in the analysis. When there are many individuals who have missing values this approach can reduce the effective sample size considerably. In some circumstances, ignoring the cases with missing values can bias an analysis. See also **available case analysis**.

Complete linkage cluster analysis An agglomerative hierarchical clustering method in which the distance between two clusters is defined as the greatest distance between a member of one cluster and a member of the other. This distance measure is illustrated in Figure 14.

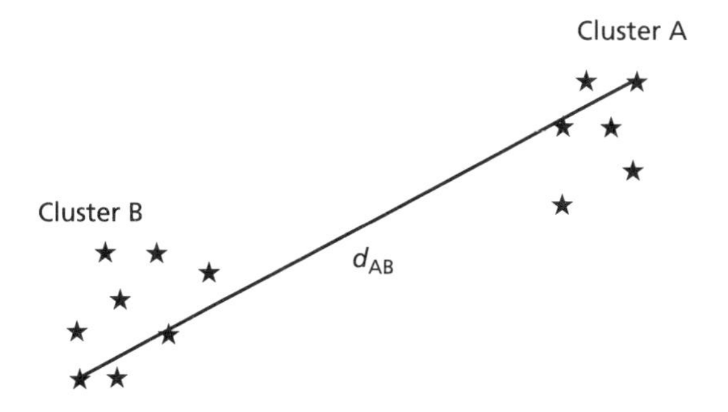

Figure 14 Between group distance used in complete linkage cluster analysis.

Completely randomized design An experimental design in which the treatments are allocated to the experimental units purely on a chance basis.

Component bar chart A bar chart that shows the component parts of the aggregate represented by the total length of the bar. The component parts are shown as sectors of the bar with lengths in proportion to their relative size. Shading or colour can be used to enhance the display. An example of such a chart involving people's perception of their health is shown in Figure 15.

Composite hypothesis A hypothesis that specifies more than a single value for a parameter. For example, the hypothesis that the mean of a population is greater than some value.

Compound symmetry The property possessed by a variance—covariance matrix of a set of multivariate data when its main diagonal elements are equal to one another, and additionally its off-diagonal elements are also equal. Consequently the matrix has the general form:

$$\Sigma = \begin{pmatrix} \sigma^2 & \rho\sigma^2 & \rho\sigma^2 & \cdots & \rho\sigma^2 \\ \rho\sigma^2 & \sigma^2 & & & \rho\sigma^2 \\ \vdots & \vdots & & & \\ \rho\sigma^2 & \rho\sigma^2 & \cdots & & \sigma^2 \end{pmatrix}$$

where ρ is the assumed common correlation coefficient of the measures. Of the most importance in the analysis of longitudinal data since it is the correlation structure assumed by the simple mixed effects model often used to analyse such data. See also **Mauchly test**.

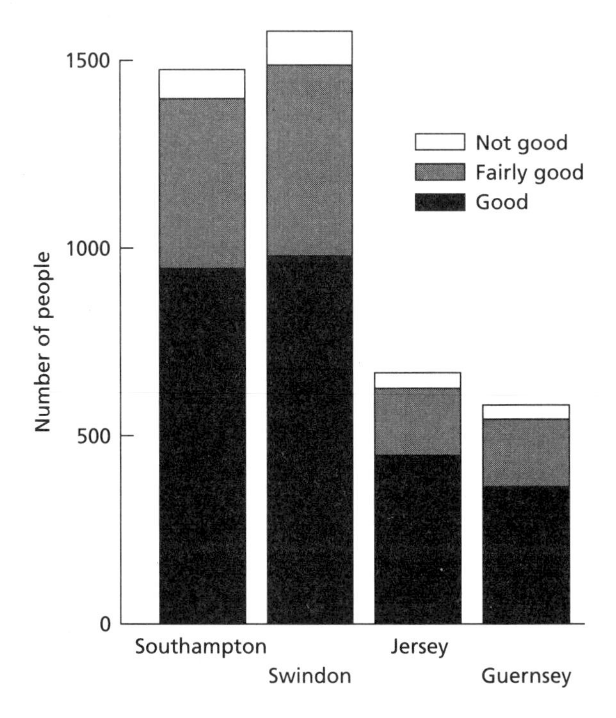

Figure 15 A component bar chart showing subjective health assessment in four regions.

Computer-aided diagnosis Computer programs designed to support clinical decision making. In general, such systems are based on the repeated application of Bayes' Theorem. In some cases, a reasoning strategy is implemented that enables the programs to conduct clinically pertinent dialogue and explain their decisions. Such programs have been developed in a variety of diagnostic areas: for example, the diagnosis of social anxiety. See also **expert system**.

Computer intensive methods Statistical methods that require almost identical computations on the data repeated many, many times. The term computer intensive is, of course, a relative quality and often the required 'intensive' computations may take only a few seconds or minutes on even a small PC. An example of such an approach is the bootstrap. See also **jackknife**.

Concomitant variables Synonym for **covariates**.

Conditional distribution The probability distribution of a random variable when the values of one or more other random variables are held fixed.

Mathematical details

As an example consider the **bivariate normal distribution** for two variables x and y. The conditional distribution of y given x is normal with mean $\mu_2 + \rho\sigma_2\sigma_1^{-1}(x - \mu)$ and variance $\sigma_2^2(1 - \rho^2)$.

Conditional probability The probability that an event A occurs given the outcome of some other event B. Usually written $\Pr(A|B)$. For example, the probability of a person being colour blind given that the person is male is about 0.1, and the corresponding probability given that the person is female is approximately 0.0001. It is not, of course, necessary that $\Pr(A|B) = \Pr(B|A)$; the probability of having spots given that a patient has measles, for example, is very high; the probability of measles given that a patient has spots is, however, much lower. If $\Pr(A|B) = \Pr(A)$ then the events A and B are said to be independent. Conditional probabilities are frequently misunderstood and the misunderstanding can have serious consequences. The tabloid press, for example, often reports DNA matching in a criminal case as 'a one in a million chance'. This is, however, the conditional probability of a match given the suspect is innocent. What is more relevant is the conditional probability of the suspect being innocent given a match. This may be quite different. For example, assuming 50 million people in the UK, 50 might be expected to match the DNA pattern found. Consequently Pr (suspect being innocent|match) = 49/50 (assuming there is a single guilty individual). See also **Bayes' Theorem**.

Confidence interval (CI) A range of values, calculated from the sample observations, that are believed, with a particular probability, to contain the true parameter value. A 95% confidence interval, for example, implies that, were the estimation process repeated again and again, then 95% of the calculated intervals would be expected to contain the true parameter value. Note that the stated probability level refers to properties of the interval and not to the parameter itself, which is not considered a **random variable** (although, see **Bayesian inference**).

Mathematical details

Formula for $100(1 - \alpha)\%$ confidence interval for difference in two means where α is the required significance level:

$$(\bar{x}_1 - \bar{x}_2) \pm t_{n;\alpha} \times s\sqrt{\frac{1}{n_1} + \frac{1}{n_2}}$$

where $\bar{x}_1$ = mean of the n_1 observations in group one and $\bar{x}_2$ = mean of the n_2 observations in group two and

$$s = \sqrt{\frac{(n_1 - 1)s_1^2 + (n_2 - 1)s_2^2}{n_1 + n_2 - 2}}$$

s_1^2 and s_2^2 are the variances of the observations in groups one and two respectively, $n = n_1 + n_2 - 2$ and $t_{n;\alpha}$ is the t-value for significance level α and n degrees of freedom.

Numerical example

IQ scores were obtained for 10 academic psychologists and 10 politicians. Means and variances were as follows:

Group	Mean	Variance
Psychologists	115	100
Politicians	105	225

The difference between sample means is 10, and the pooled estimate of the standard deviation, s, is given by:

$$s = \sqrt{\frac{9 \times 100 + 9 \times 225}{23}} = 11.28$$

The t-value for $\alpha = 0.05$ and 23 degrees of freedom is 2.069. The required 95% confidence interval for the mean difference is therefore given by:

$$10 \pm 2.069 \times 11.28 \times \sqrt{\frac{1}{10} + \frac{1}{10}} = (-0.43, 50.43)$$

Since this interval contains the value zero these data give no convincing evidence of a difference in mean IQ score of the two populations, although the proximity of the lower limit to zero might be suggestive of such a difference.

Confirmatory data analysis A term often used for model fitting and inferential statistical procedures to distinguish them from the methods of **exploratory data analysis**.

Confirmatory factor analysis (CFA) See **factor analysis**.

Confounding A process observed in some **factorial designs** in which it is impossible to differentiate between some **main effects** or **interactions**, on the basis of the particular design used. In essence, the **contrast** that measures one of the effects is exactly the same as the contrast that measures the other. The two effects are usually referred to as *aliases*.

Conservative and non-conservative tests Terms usually encountered in discussions of **multiple comparison tests**. Non-conservative tests provide poor control over the **per-experiment error rate**. Conservative tests, on the other hand, may limit the **per-comparison error rate** to unnecessarily low values, and tend to have low **power** unless the sample size is large.

Consistency checks Checks built into the collection of a set of observations to assess their internal consistency. For example, data on age might be collected directly and also by asking about date of birth.

Contingency coefficient A measure of association, C, of the two variables forming a **two-dimensional contingency table**.

Mathematical details

The coefficient is given by

$$C = \sqrt{\frac{\chi^2}{\chi^2 + N}}$$

where χ^2 is the usual chi-squared statistic for testing the independence of the two variables and N is the sample size. The coefficient lies between 0 and 1 and attains its lower limit in the case of complete independence: that is, when $\chi^2 = 0$. In general, however, it cannot attain the upper limit of 1.

Numerical example

The data given in the 2×2 table below were collected during an investigation of the relationship between smoking and age.

		Under 40	**Over 40**	**Total**
Amount of smoking	<20/day	50	15	65
	>20/day	10	25	35
Total		60	40	100

The contingency coefficient is given by

$$C = \sqrt{\frac{22.16}{22.16 + 100}} = 0.43$$

See also **phi-coefficient**.

Contingency tables The tables arising when observations on a number of categorical variables are cross-classified. Entries in each cell are the number of individuals with the corresponding combination of variable values. Most common are tables involving two categorical variables known as two-dimensional contingency tables. The analysis of such two-dimensional tables generally involves testing for the independence of the two variables using the familiar chi-squared statistic. Three- and higher-dimensional tables are now routinely analysed using log-linear models.

Numerical example

Retarded activity amongst psychiatric patients

	Affective psychosis	**Schizophrenia**	**Neurosis**	**Total**
Retarded activity	12	13	5	30
No retarded activity	18	17	25	60
Total	30	30	30	90

Under the hypothesis of independence the estimated expected value in a cell is (row total × column total)/(sample size).

Table of expected values

	Affective psychosis	Schizoprehenia	Neurosis	Total
Retarded activity	10	10	10	30
No retarded activity	20	20	20	60
Total	30	30	30	90

$$\chi^2 = (12-10)^2/10 + (13-10)^2/10 + (5-10)^2/10 + (18-20)^2/20 + (17-20)^2/20 + (25-20)^2/20 = 5.7$$

Using a **chi-squared distribution** with two degrees of freedom this has an associated **P-value** of 0.06. There is evidence of an association between the two variables, suggesting here that patients with psychosis (schizophrenia and affective psychosis) show more retarded activity that those with neurosis.

Continuity correction Synonym for **Yates' correction**.

Continuous variable A measurement not restricted to particular values except in so far as this is restricted by the accuracy of the measuring instrument. Common examples include reaction time, neuroticism, and IQ. For such a variable, equal sized differences on different parts of the scale are equivalent. See also **categorical variable** and **ordinal variable**.

Contour plot A topographical map drawn from data involving observations on three variables. One variable is represented on the horizontal axis and a second variable is represented on the vertical axis. The third variable is represented by isolines (lines of constant value). These plots are often helpful in data analysis, especially when searching for maxima or minima in such data. The plots are most often used to display graphically **bivariate distributions** in which case the contours give values of the bivariate density of the two variables. An alternative method of displaying the same material is provided by the *perspective plot* in which the values of the third variable are represented by a series of lines constructed to give a three-dimensional view of the data. Figure 16 gives an example of the estimated probability density of birth and death rates for a number of countries.

Contrast A **linear function** of parameters or statistics in which the coefficients sum to zero. Most often encountered in the context of **analysis of variance**. For example, in an application involving, say, three treatment groups (with means $\bar{x}_{T_1}$, $\bar{x}_{T_2}$ and $\bar{x}_{T_3}$) and a control group (with mean $\bar{x}_C$), the following is the contrast for comparing the mean of the control group with the average of the treatment groups:

$$\bar{x}_C - \tfrac{1}{3}\bar{x}_{T_1} - \tfrac{1}{3}\bar{x}_{T_2} - \tfrac{1}{3}\bar{x}_{T_3}$$

See also **Helmert contrast** and **orthogonal contrast**.

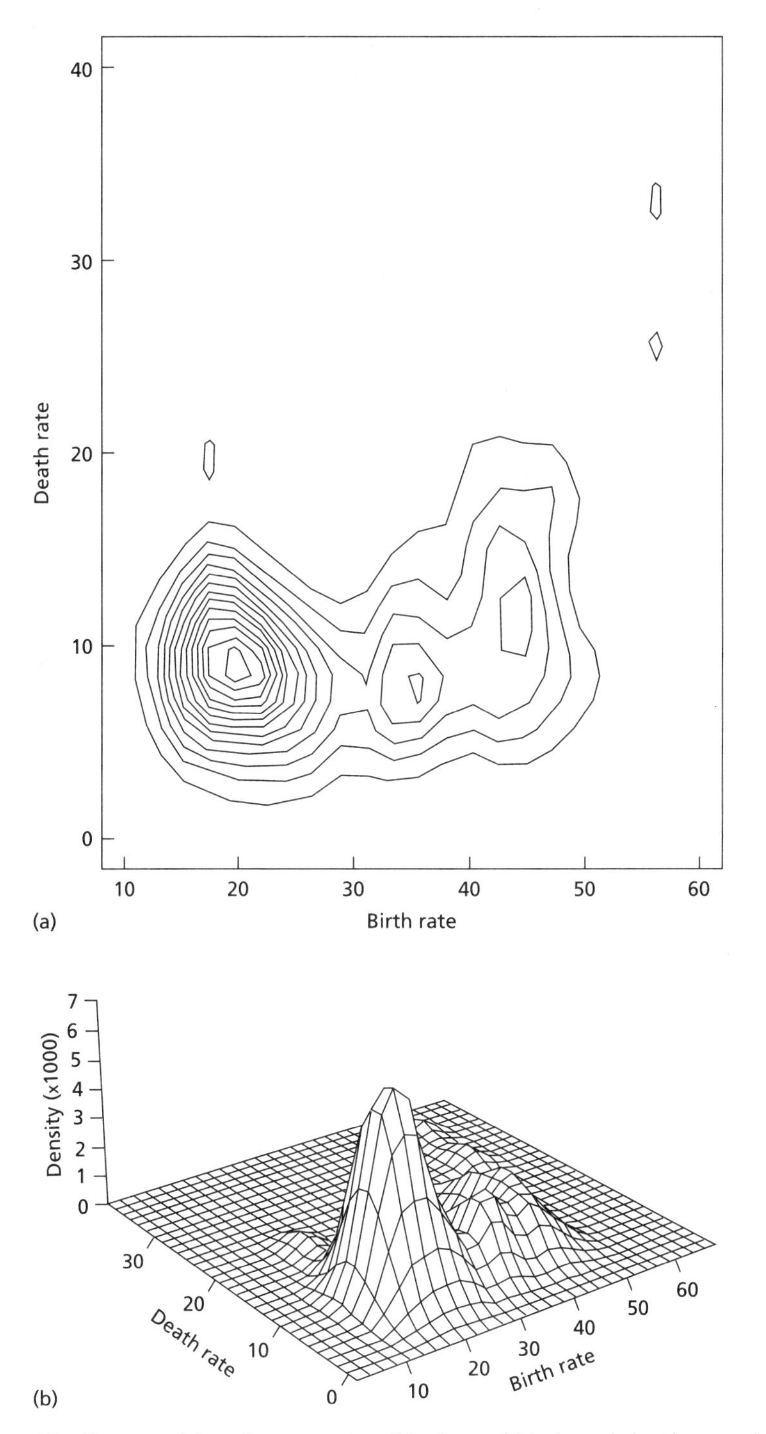

Figure 16 Contour (a) and perspective (b) plots of birth and death rates in a number of countries.

Control group In **experimental studies**, a collection of individuals to which the experimental procedure of interest is not applied. In **observational studies**, most often used for a collection of individuals not subjected to the **risk factor** under investigation. In many studies, the controls are drawn from the same clinical source as the cases, to ensure that they represent the same catchment population and are subject to the same selective factors. These would be termed *hospital controls*. An alternative is to use controls taken from the population from which the cases are drawn (*community controls*). The latter is suitable only if the source population is well defined and the cases are representative of the cases in this population.

Correlated samples *t*-test Synonym for **matched pairs *t*-test**.

Correlation A general term for interdependence between pairs of variables. See also **association**.

Correlation coefficient An index that quantifies the linear relationship between a pair of variables. In a **bivariate normal distribution**, for example, the parameter ρ.

Mathematical details

The population correlation, ρ, is usually estimated by *Pearson's product moment correlation coefficient*, r, given by

$$r = \frac{\sum_{i=1}^{n}(x_i - \bar{x})(y_i - \bar{y})}{\sqrt{\sum_{i=1}^{n}(x_i - \bar{x})^2(y_i - \bar{y})^2}}$$

where $(x_1, y_1), (x_2, y_2), \ldots, (x_n, y_n)$ are the n sample values of the two variables of interest. The coefficient takes values between -1 and 1, with the sign indicating the direction of the relationship and the numerical magnitude its strength. Values of -1 or 1 indicate that the sample values fall on a straight line. A value of zero indicates the lack of any linear relationship between the two variables. Figure 17 shows a number of scatter diagrams and their associated correlation coefficients.

See also **Fisher's *z* transformation, Spearman's rho, intra-class correlation** and **Kendall's tau statistics**.

Correlation matrix A square, **symmetric matrix** with rows and columns corresponding to variables, in which the off-diagonal elements are correlations between pairs of variables, and elements on the main diagonal are unity.

Mathematical details

For a set of q variables the correlation matrix $\mathbf{R}$ has the form

$$\mathbf{R} = \begin{bmatrix} 1 & r_{12} & \cdots & r_{1q} \\ r_{21} & \cdots & \cdots & \cdots \\ \vdots & \vdots & \vdots & \vdots \\ r_{q1} & \cdots & \cdots & 1 \end{bmatrix}$$

where r_{ij} is the correlation coefficient for variables i and j.

Numerical example

A sample of 1500 students in the USA were asked about their usage of certain substances. Usage was rated on a five-point scale from never to frequent. The following is the observed correlation matrix.

Correlations between usage rates of five substances amongst students in the USA

	Cigarettes	Beer	Wine	Liquor	Cocaine
Cigarettes	1.00	0.45	0.42	0.44	0.11
Beer	0.45	1.00	0.62	0.60	0.07
Wine	0.42	0.62	1.00	0.58	0.05
Liquor	0.44	0.60	0.58	1.00	0.12
Cocaine	0.11	0.07	0.05	0.12	1.00

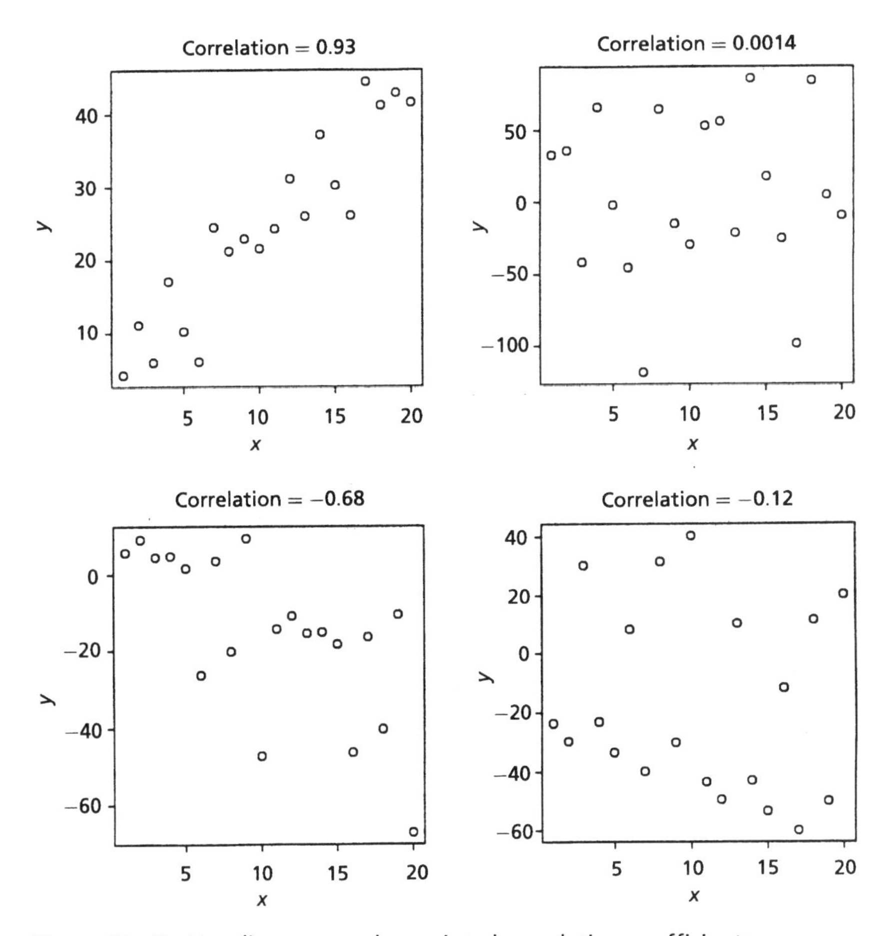

Figure 17 Scatter diagrams and associated correlation coefficients.

Correlogram See **autocorrelation**.

Correspondence analysis A method for deriving a set of coordinate values representing the row and column categories of a **contingency table**, and thus allowing the associations in the table to be displayed graphically. The derived coordinates are analogous to those resulting from a **principal components analysis**, except that they involve a partition of a **chi-squared statistic** rather than the total variance. Figure 18 shows a plot of the coordinates derived from the following contingency table ('E' indicates an eye colour and 'h' a hair colour). [*Developmental Brain Dysfunction*, 1997, 10, 28–39].

	Hair colour				
Eye colour	**Fair**	**Red**	**Medium**	**Dark**	**Black**
Light	688	116	584	188	4
Blue	326	38	241	110	3
Medium	343	84	909	412	26
Dark	98	48	403	681	81

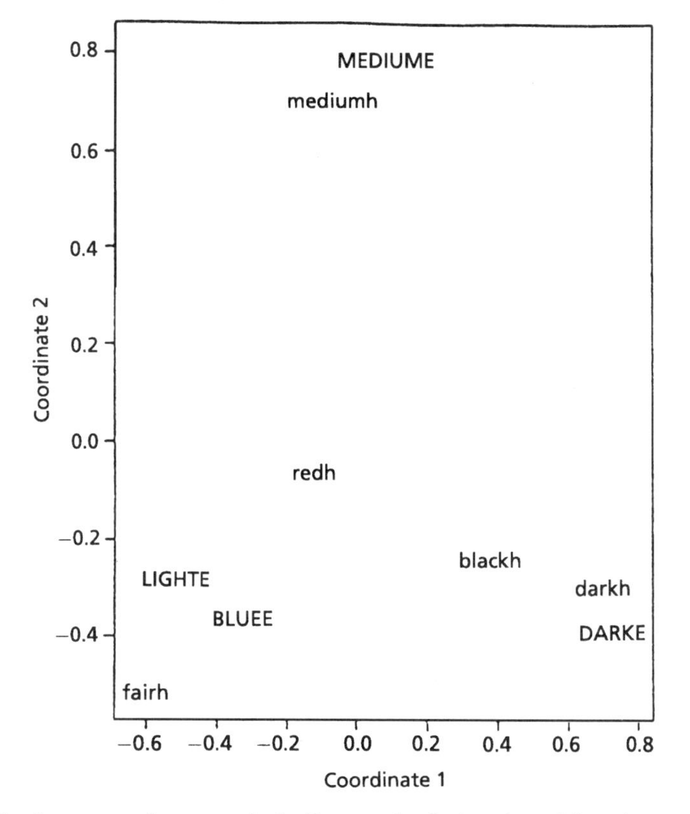

Figure 18 Correspondence analysis diagram for hair colour (h) and eye colour (E) data.

Count data Data obtained by counting the number of occurrences of a particular event, rather than by taking measurements on some scale.

Covariance The **expected value** of the product of the deviations of two **random variables**, x and y, from their respective means, μ_x and μ_y.

Mathematical details

Covariance is defined as

$$\text{Cov}(x, y) = E(x - \mu_x)(y - \mu_y)$$

The corresponding sample statistic is

$$c_{xy} = \frac{1}{n}\sum_{i=1}^{n}(x_i - \bar{x})(y_i - \bar{y})$$

where (x_i, y_i), $i = 1, \ldots, n$ are the sample values on the two variables, and $\bar{x}$ and $\bar{y}$ their respective means.

For a numerical example see **variance–covariance matrix** entry.

See also **correlation coefficient**.

Covariance matrix See **variance–covariance matrix**.

Covariance modelling Synonymous with **structural equation modelling**.

Covariates Often used simply as an alternative name for explanatory variables, but perhaps more specifically to refer to variables that are not of primary interest in an investigation, but are measured because it is believed that they are likely to affect the response variable, and consequently need to be included in analyses and model building.

Cox's proportional hazards model A method used for investigating the relationship between **survival times** and a set of explanatory variables. The survival times are not modelled directly but via their **hazard function**.

Mathematical details

The model involved is:

$$\ln h(t) = \ln h_0(t) + \beta_1 x_1 + \beta_2 x_2 + \cdots + \beta_q x_q$$

where $x_1, x_2, \ldots, x_q$ are the explanatory variables of interest, and $h(t)$ is the hazard function. The function $h_0(t)$ is the baseline hazard function and is an arbitrary function of time. For any two individuals at any point in time, the ratio of the hazard functions is a constant. Estimates of the parameters, $\beta_1, \beta_2, \ldots, \beta_q$, are obtained by **maximum likelihood estimation** and depend only on the order in which events occur, not on the exact times of their occurrence.

Numerical example

The 'survival data' for this example are the times that heroin addicts remained in a clinic for methadone maintenance treatment. Here the endpoint of interest is not death as the word 'survival' suggests, but termination of treatment. Some subjects were still in the clinic at the time these data were recorded and this is indicated by the status variable which is equal to one if the addict had departed and zero otherwise. Two explanatory variables are of interest: maximum methadone dose and whether or not the addict had a prison record. The data of 15 subjects were as follows:

Subject	Status	Time in clinic	Prison	Dose
1	1	428	0	50
2	1	275	1	55
3	1	262	0	55
4	1	183	0	30
5	1	259	1	65
6	1	714	0	55
7	1	438	1	65
8	0	796	1	60
9	1	892	0	50
10	1	393	1	65
11	0	161	1	80
12	1	836	1	60
13	1	523	0	55
14	1	612	0	70
15	1	212	1	60

The estimated coefficients from Cox's proportional hazards model and their standard errors are as follows:

Variable	Estimated coefficient	SE
Prison	0.4505	0.7138
Dose	−0.0413	0.0566

Comparing each estimated coefficient with its standard error indicates that neither differs from zero (the ratios lie within the range $-2, 2$). Consequently neither is associated with time spent in clinic.

[*Quality of Life Research: An International Journal of Quality of Life Aspects of Treatment, Care and Rehabilitation*, 1997, 6, 151–158].

Cramér's *V* A measure of association for the two variables forming a **two-dimensional contingency table**. Related to the **phi-coefficient**, ϕ, but can be applied to tables larger than 2×2.

Mathematical details

The coefficient is given by:

$$V = \sqrt{\left\{\frac{\phi}{\min[(r-1)(c-1)]}\right\}}$$

where r is the number of rows of the table and c is the number of columns.

See also **contingency coefficient**. [*Behaviour Modification*, 1996, 20, 281–299].

Critical region The values of a **test statistic** that lead to rejection of a null hypothesis. Suppose, for example, a **z-test** is being used to test that the mean of a population is 10 against the alternative hypothesis that it is not 10. If the significance level chosen is 0.05, the critical region consists of values of z less than -1.96 and greater than $+1.96$. See also **acceptance region**.

Critical value The value(s) with which a statistic calculated from sample data is compared in order to decide whether a null hypothesis should be rejected. The value is related to the particular significance level chosen. For example, when using a two-tailed **z-test** with significance level 0.05, the values are -1.96 and 1.96.

Cronbach's alpha An index of the internal consistency of a psychological test. If the test consists of n items and an individual's score is the total answered correctly, then the coefficient is given specifically by

$$\alpha = \frac{n}{n-1}\left[1 - \frac{1}{\sigma^2}\sum_{i=1}^{n}\sigma_i^2\right]$$

where σ^2 is the variance of the total scores and σ_i^2 is the variance of the set of 0, 1 scores representing correct and incorrect answers on item i. If α is large then it can be assumed that the total score is reasonably representative of the individual scores on the test. [*Psychological Assessment*, 1998, 10, 64–70].

Crossover design A type of **longitudinal study** in which participants receive different procedures on different occasions. Random allocation is used to determine the order in which the procedures are received. The simplest such design involves two groups of participants: one receives each of two procedures, A and B, in order AB, while the other receives them in the reverse order. This is known as a *two-by-two crossover design*. Since the procedure comparison is 'within participant' rather than 'between participant', it is likely to require fewer participants to achieve a given **power**. The analysis of such designs is not necessarily straightforward, because of the possibility of *carryover effects*: that is, residual effects of the procedure received on the first occasion that remain present into the second occasion. An attempt to minimize this problem is often made by including a **wash-out period** between the two procedure occasions. Some authorities have suggested that this type of design should only be used if such carryover effects can be ruled out *a priori*. [*British Journal of Clinical Psychology*, 1998, 37, 69–82].

Cross-sectional study A study not involving the passing of time. All information is collected at the same time and participants are contacted only once. Many surveys are of this type. The temporal sequence of cause and effect cannot be addressed in such a study, but it may be suggestive of an association that should be investigated more fully by, for example, a **prospective study**.

Cross-validation The division of data into two approximately equal sized subsets, one of which is used to estimate the parameters in some model of interest, and the other is used to assess whether the model with these parameter values fits adequately. See also **jackknife**.

Cumulative probability distribution See **probability distribution**.

Curvilinear effects A general name for polynomial terms in **regression analysis**.

D

Data dredging A term used to describe comparisons made within a data set not specifically prescribed prior to the start of the study.

Data editing The action of removing format errors and keying errors from data.

Data matrix See **multivariate data**.

Data mining A term once used in a pejorative sense for the process of considering a large number of models, including many which are 'data-driven', in order to obtain a good fit. In recent years, however, the process has become more statistically respectable. See also **data dredging**.

Data reduction The process of summarizing large amounts of data by forming frequency distributions, histograms, scatter diagrams, etc., and calculating statistics such as means, variances and correlation coefficients. The term is also used when obtaining a low-dimensional representation of multivariate data by procedures such as principal components and factor analysis.

Data screening The initial assessment of a set of observations to see whether or not they appear to satisfy the assumptions of the methods to be used in their analysis. Techniques which highlight possible outliers, or, for example, departures from normality are important in this phase of an investigation. See also **initial data analysis**.

Data set A general term for observations and measurements collected during any type of statistical investigation.

Degrees of freedom (df) An elusive concept that occurs throughout statistics. Essentially, the term means the number of independent units of information in a sample relevant to the estimation of a parameter or calculation of a statistic. For example, in a two-by-two contingency table with a given set of marginal totals, only one of the four cell frequencies is free, and the table therefore has a single degree of freedom. In many cases the term corresponds to the number of parameters in a model. Also used to refer to a parameter of various families of distributions: for example, Student's t-distribution and the F-distribution.

Dendrogram A term usually encountered in the application of agglomerative hierarchical clustering methods, where it refers to the 'tree-like' diagram illustrating the series of steps taken by the method in proceeding from n single-member 'clusters' to a single group containing all n individuals.

Numerical example

The application of **single linkage clustering** to the following matrix of **Euclidean distances** between five points

$$\mathbf{D} = \begin{pmatrix} 0.0 & & & & \\ 2.0 & 0.0 & & & \\ 6.0 & 5.0 & 0.0 & & \\ 10.0 & 9.0 & 4.0 & 0.0 & \\ 9.0 & 8.0 & 5.0 & 3.0 & 0.0 \end{pmatrix}$$

gives the dendrogram shown in Figure 19 (details of the calculations are given in the **single linkage clustering** entry).

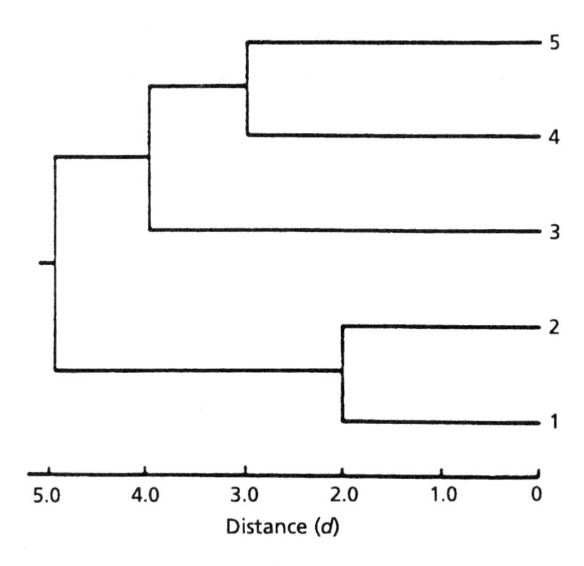

Figure 19 Single linkage clustering dendrogram.

Dependent variable The variable of primary importance in psychological investigations, since the major objective is usually to study the effects of different conditions and/or other explanatory variables on this variable, and to provide suitable models for its relationship with the explanatory variables.

Design matrix A matrix that occurs in **regression analysis** and specifies the relationships between the dependent variable and the explanatory variable.

Mathematical details

Consider a regression model for five individuals which involves a dependent variable y and three explanatory variables x_1, x_2, x_3. The model for all five individuals can be written as

$$y_1 = \beta_0 + \beta_1 x_{11} + \beta_2 x_{21} + \beta_3 x_{31} + \varepsilon_1$$
$$y_2 = \beta_0 + \beta_1 x_{12} + \beta_2 x_{22} + \beta_3 x_{32} + \varepsilon_2$$
$$y_3 = \beta_0 + \beta_1 x_{13} + \beta_2 x_{23} + \beta_3 x_{33} + \varepsilon_3$$
$$y_4 = \beta_0 + \beta_1 x_{14} + \beta_2 x_{24} + \beta_3 x_{34} + \varepsilon_4$$
$$y_5 = \beta_0 + \beta_1 x_{15} + \beta_2 x_{25} + \beta_3 x_{35} + \varepsilon_5$$

Here the design matrix, **X**, is given by

$$\mathbf{X} = \begin{bmatrix} 1 & x_{11} & x_{21} & x_{31} \\ 1 & x_{12} & x_{22} & x_{32} \\ 1 & x_{13} & x_{23} & x_{33} \\ 1 & x_{14} & x_{24} & x_{34} \\ 1 & x_{15} & x_{25} & x_{35} \end{bmatrix}$$

enabling the model to be written more easily as

$$y = \mathbf{X}\boldsymbol{\beta} + \boldsymbol{\varepsilon}$$

where

$$y' = [y_1, y_2, y_3, y_4, y_5], \quad \boldsymbol{\beta}' = [\beta_0, \beta_1, \beta_2, \beta_3] \quad \text{and} \quad \boldsymbol{\varepsilon}' = [\varepsilon_1, \varepsilon_2, \varepsilon_3, \varepsilon_4, \varepsilon_5]$$

Determinant A value associated with a **square matrix** calculated from the elements of the matrix. Arises in accounts of many methods of **multivariate analysis**.

Example

For the matrix **A** given by

$$\mathbf{A} = \begin{bmatrix} a & b \\ c & d \end{bmatrix}$$

the determinant of **A**, det(**A**), is given by

$$ad - bc$$

Deviance A measure of the extent to which a particular model differs from the **saturated model** for a data set. Defined explicitly in terms of the **likelihoods** of the two models as

$$D = -2[\ln L_c - \ln L_s]$$

where L_c and L_s are the likelihoods of the current model and the saturated model, respectively. Large values of D are encountered when L_c is small relative to L_s, indicating that the current model is a poor one. Small values of D are obtained in the reverse case. The deviance has, asymptotically, a **chi-squared distribution** with degrees of freedom equal to the difference in the number of parameters in the two models. See also **likelihood ratio**.

DF (df) Abbreviation for **degrees of freedom**.

Diagonal matrix A square matrix whose off-diagonal elements are all zero. For example:

$$\mathbf{D} = \begin{bmatrix} 10 & 0 & 0 \\ 0 & 5 & 0 \\ 0 & 0 & 3 \end{bmatrix}$$

Such matrices are often encountered in multivariate analysis.

Dichotomous variable Synonym for **binary variable**.

Digit preference The personal and often subconscious bias that frequently occurs in the recording of observations. Usually most obvious in the final recorded digit of a measurement. The phenomenon is illustrated in Figure 20.

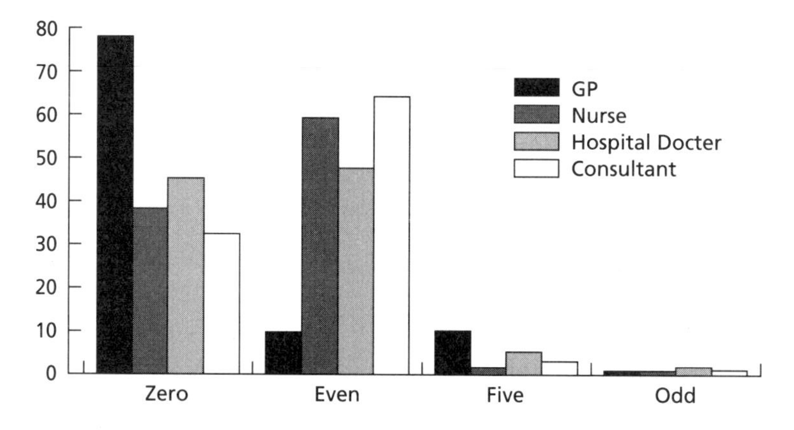

Figure 20 Some empirical evidence of digit preference.

Disattentuated correlation A correlation between two variables after correction for errors of measurement. See also **attenuation**.

Discrete variables Variables having only integer values: for example, number of trials needed by a rat to learn to run a maze.

Discriminant analysis A class of techniques that are concerned with deriving rules for classifying an object or individual into one of a number of predefined groups, using the information provided by a number of variable values recorded on the object or individual. Some possible applications are:

- determining what category of vulnerability risk (e.g. low risk or high risk) an individual falls into,
- predicting the success or failure of a new product,
- classifying mentally ill patients into one of a number of diagnostic categories.

The most commonly occurring situation is that involving just two groups, where the most widely applied method is Fisher's linear discriminant function. [*British Journal of Clinical Psychology*, 1997, 36, 627–629].

Dispersion The amount by which a set of observations deviate from their mean. When the values of a set of observations are close to their mean, the dispersion is less than when they are spread out widely from the mean. See also **variance**.

Dissimilarity coefficient Used for a judgement of the dissimilarity of two stimuli by an observer or for a coefficient assessing the difference between two stimuli calculated from a set of variable values available for each. See also **dissimilarity matrix**.

Dissimilarity matrix A square, **symmetric matrix** whose elements are **dissimilarity coefficients**.

Mathematical details

A dissimilarity matrix **D** has the form

$$\mathbf{D} = \begin{bmatrix} 0 & \delta_{12} & \cdots & \delta_{1N} \\ \delta_{21} & & & \\ \vdots & & & \\ \delta_{N1} & & & 0 \end{bmatrix}$$

where N is the number of stimuli and δ_{ij} is the dissimilarity coefficient for stimuli i and j, and $\delta_{ij} = \delta_{ji}$.

Numerical example

(1) Dissimilarity matrix calculated from variable values.

Consider the following set of data consisting of four individuals each having observations on two variables:

Individual	Memory for word list after 5 s delay	Memory for word list after 10 min delay
1	5	6
2	10	3
3	1	2
4	8	7

Define a measure of dissimilarity for a pair of individuals as the absolute value of the greatest difference on their two variable values. Then the dissimilarity matrix for the individuals is **D**

$$\mathbf{D} = \begin{bmatrix} 0 & 5 & 4 & 3 \\ 5 & 0 & 9 & 4 \\ 4 & 9 & 0 & 7 \\ 3 & 4 & 7 & 0 \end{bmatrix}$$

(2) Directly observed dissimilarity matrix.

As an example of a directly observed dissimilarity matrix, the table below shows the number of times 15 Congressmen from New Jersey voted differently in the House of Representatives on 19 environmental bills (R = Republican, D = Democrat):

Name (party)	1	2	3	4	5	6	7	8	9	10	11	12	13	14	15
1. Hunt (R)	0														
2. Sandman (R)	8	0													
3. Howard (D)	15	17	0												
4. Thompson (D)	15	12	9	0											
5. Frelinghuysen (R)	10	13	16	14	0										
6. Forsythe (R)	9	13	12	12	8	0									
7. Widnall (R)	7	12	15	13	9	7	0								
8. Roe (D)	15	16	5	10	13	12	17	0							
9. Heltoski (D)	16	17	5	8	14	11	16	4	0						
10. Rodino (D)	14	15	6	8	12	10	15	5	3	0					
11. Minish (D)	15	16	5	8	12	9	14	5	2	1	0				
12. Rinaldo (D)	16	17	4	6	12	10	15	3	1	2	1	0			
13. Maraziti (R)	7	13	11	15	10	6	10	12	13	11	12	12	0		
14. Daniels (D)	11	12	10	10	11	6	11	7	7	4	5	6	9	0	
15. Patten (D)	13	16	7	7	11	10	13	6	5	6	5	4	13	9	0

Such a matrix might be used for **multidimensional scaling** in order to explore allegiances etc. amongst the Congressmen.

Distribution-free methods Statistical methods which do not assume a particular form for the **probability distribution** of the observations. Consequently the techniques are valid under relatively general assumptions about the underlying population. Often such methods involve only the ranks of the observations rather than the observations themselves. Examples are **Wilcoxon's signed rank test** and **Friedman's two-way analysis of variance**. In many cases these tests are only marginally less powerful than their analogues which assume a particular population distribution (usually a **normal distribution**), even when that assumption is true. Also commonly known as *non-parametric methods,* although the terms are not completely synonymous.

Dot plot A more effective display than a number of other methods, for example, **pie charts** and **bar charts**, for displaying quantitative data which are labelled. Figure 21 gives an example.

Double blind See **blinding**.

Draughtsman's plot An arrangement of the pairwise **scatter diagrams** of the variables in a set of **multivariate data** in the form of a grid or matrix with shared scales. Such an arrangement may be extremely useful in the initial examination of the data. The example in Figure 22 involves plots of rates for different types of crime in the USA. In the six graphs in the bottom row the vertical scale is the car theft rate and the six horizontal scales are the murder, rape, robbery, assault, burglary and theft rates. The upper left-hand triangle contains all 21 pairs of

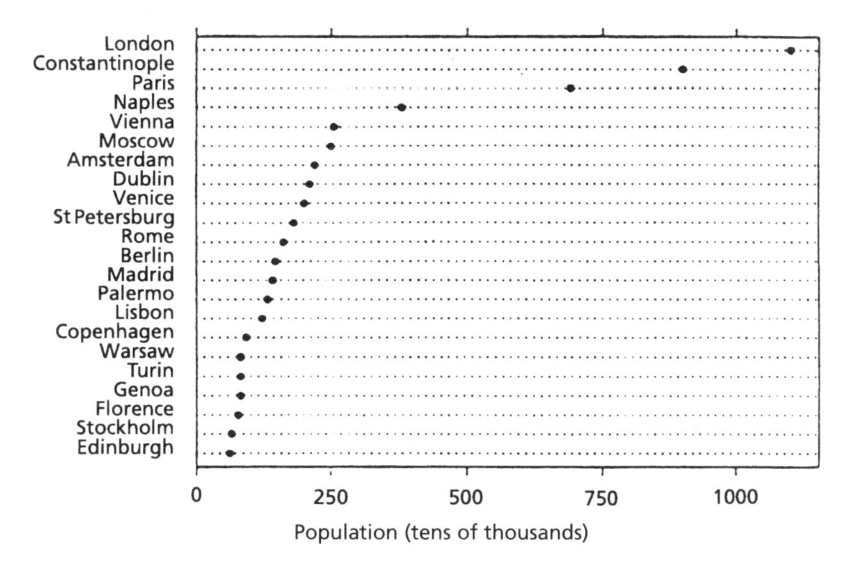

Figure 21 Dot plot of population size of a number of cities.

scatter plots, and so does the lower right-hand triangle. The reason for including both the upper and lower triangles in the matrix, despite the seeming redundancy, is that it enables a row and column to be visibly scanned to see one variable against all others, with the scales for the one variable lined up along the horizontal or the vertical.

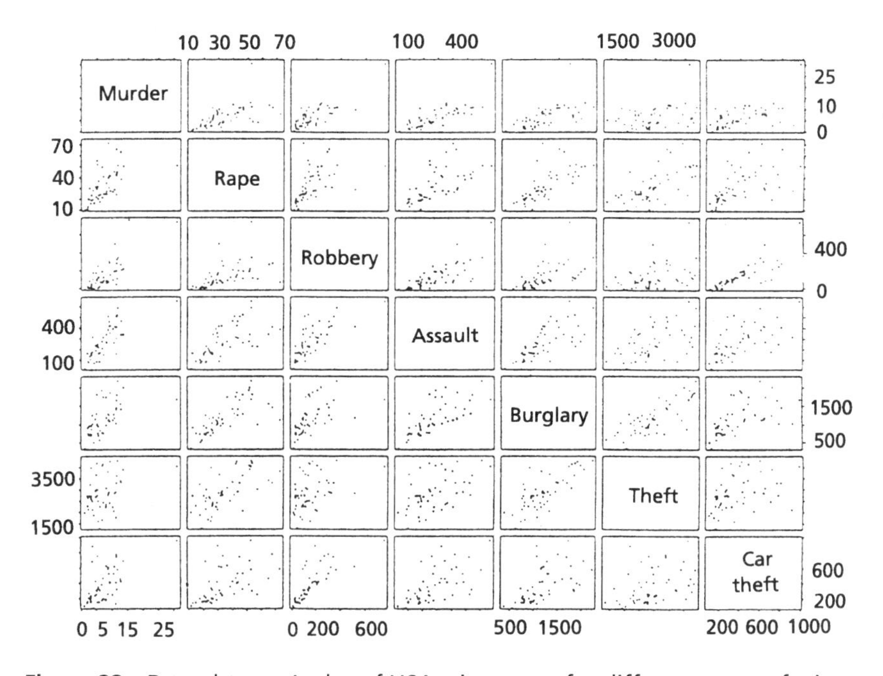

Figure 22 Draughtsman's plot of USA crime rates for different types of crime.

Dual scaling Synonym for **correspondence analysis**.

Dummy variables The variables resulting from recoding **categorical variables** with more than two categories into a series of **binary variables**. Marital status, for example, if originally labelled 1 for married, 2 for single and 3 for divorced, widowed or separated, could be redefined in terms of two variables as follows:

- Variable 1: 1 if single, 0 otherwise
- Variable 2: 1 if divorced, widowed or separated, 0 otherwise.

For a married person, both new variables would be zero. In general a categorical variable with k categories would be recoded in terms of $k - 1$ dummy variables. Such recoding is used before **polychotomous variables** are used as explanatory variables in a **regression analysis**, to avoid the unreasonable assumption that the original numerical codes for the categories, i.e. the values $1, 2, \ldots, k$, correspond to an **interval scale**.

Duncan's multiple range test A modified form of the **Newman–Keuls test**.

Durbin–Watson test A test of the assumption usually made in **linear regression** and **multiple regression** that the error terms in the model are independent of one another.

Mathematical details

The test statistic is

$$D = \frac{\sum_{i=2}^{n} (r_i - r_{i-1})^2}{\sum_{i=1}^{n} r_i^2}$$

where $r_i = y_i - \hat{y}_i$, and y_i and $\hat{y}_i$ are, respectively, the observed and predicted values of the response variable for individual i. D becomes smaller as the **serial correlations** increase. **Critical values** are available in many standard statistical tables.

Dynamic graphics Computer graphics for the exploration of multivariate data which allow the observations to be rotated and viewed from all directions. Particular sets of observations can be highlighted. Often useful for discovering structure or pattern in the data: for example, the presence of distinct clusters of observations.

E

E Abbreviation for **expected value**.

EDA Abbreviation for **exploratory data analysis**.

Effect Generally used for the change in a dependent variable associated with a change in one or more explanatory or factor variables.

Effective sample size The sample size after dropouts, deaths and other specified exclusions from the original sample.

Effect size A general term for the estimated size of effects like the difference in treatment groups on some scale.

Eigenvalues A term often encountered in accounts of methods for multivariate analysis. Mathematically the roots of a certain polynomial equation, which give the variances of particular linear functions of the observed variables. See also **eigenvector** and **principal components analysis**.

Eigenvector A term often encountered in accounts of methods for multivariate analysis. The elements of such vectors define linear functions of the observed variables with particular properties, for example, maximal variance. See also **eigenvalue** and **principal components analysis**.

Embase Range of databases on various aspects of medicine. Useful for literature searches. The *CD of Psychiatry* covers psychology, addiction and other aspects of behaviour. Published by Elsevier Science BV.

Empirical Based on observation or experiment rather than deduction from basic laws or theory.

End-aversion bias A term which refers to the reluctance of some people to use the extreme categories of a scale. See also **acquiescence bias**.

EQS A software package for fitting structural equation models. See also **LISREL**.

Error mean square See **mean squares**.

Error rate The proportion of subjects misclassified by a classification rule derived from a discriminant analysis.

Errors-in-variables problem See **regression dilution**.

Errors of the third kind Giving the right answer to the wrong question! (Not to be confused with Type III error.)

Estimation The process of providing a numerical value for a population parameter on the basis of information collected from a sample. If a single figure is calculated for the unknown parameter, the process is called *point estimation.* If an interval is calculated within which the parameter is likely to fall, then the procedure is called *interval estimation.* If, for example, a sample of 10 individuals from some population of interest had reaction times in seconds of 1.83, 1.79, 1.65, 1.91, 1.42, 1.30, 1.74, 1.69, 1.65, 1.41, a possible point estimate of the population mean would be the average of the 10 values, i.e. 1.64. If we calculate the standard deviation of the 10 values, a possible interval estimate is

$$(\text{mean} - 2 \times \text{sd}, \text{mean} + 2 \times \text{sd})$$

i.e. (1.22, 1.68). See also **least squares estimation**, **maximum likelihood estimation** and **confidence interval**.

Euclidean distance For two observations $x' = [x_1, x_2, \ldots, x_q]$ and $y' = [y_1, y_2, \ldots, y_q]$ from a set of **multivariate data**, the distance measure given by

$$d_{xy} = \sqrt{\sum_{i=1}^{q}(x_i - y_i)^2}$$

Often used as the basis of a **multidimensional scaling** or **cluster analysis** of the data.

Evidence-based medicine Described by its leading proponent as 'the conscientious, explicit, and judicious use of current best evidence in making decisions about the care of individual patients, and integrating individual clinical experience with the best available external clinical evidence from systematic research'. See also **Cochrane collaboration**.

Expected frequencies A term usually encountered in the analysis of **contingency tables**. Such frequencies are estimates of the values to be expected under the hypothesis of interest. In a two-dimensional table, for example, the values under independence are calculated from the product of the appropriate row and column totals divided by the total number of observations.

Expected value (E) The mean of the **probability distribution** of a **random variable**.

Mathematical details

For a discrete random variable with probability distribution $P(x)$,

$$E(x) = \sum_{x} x\, P(x)$$

For a continuous random variable with probability distribution $f(x)$,

$$E(x) = \int_{x} x f(x)\, \mathrm{d}x$$

Example

Consider a random variable with an exponential distribution, i.e. $f(x) = \lambda e^{-\lambda x}$

$$E(x) = \int_0^\infty x \lambda e^{-\lambda x} dx = \frac{1}{\lambda}$$

Experimental design The arrangement and procedures used in an experimental study. Some general principles of good design are: simplicity, avoidance of bias, the use of random allocation for forming comparison groups, replications and adequate sample size.

Experimental hypothesis Synonymous with **alternative hypothesis**.

Experimental study A study in which conditions, treatments, etc. can be manipulated by the investigator.

Experimentwise error rate Synonym for **per-experiment error rate**.

Expert systems Computer programs designed to mimic the role of an expert human consultant. Such systems are able to cope with complex problems because of their ability to manipulate symbolic, rather than just numeric, information, and their use of judgmental or heuristic knowledge to construct intelligible solutions to problems. Most often employed in a clinical setting where well-known examples include the MYCIN system, developed at Stanford University, ABEL, developed at MIT, and Sleep-EVAL. See also **computer-aided diagnosis**. [*Neuropsychological Rehabilitation*, 1997, 7, 419–439].

Exploratory data analysis (EDA) An approach to data analysis that emphasizes the use of informal graphical procedures, not based on prior assumptions about the structure of the data or on formal models for the data. The essence of this approach is that, broadly speaking, data are assumed to possess the following structure:

$$\text{Data} = \text{Smooth} + \text{Rough}$$

where the 'Smooth' is the underlying regularity or pattern in the data. The objective of the exploratory approach is to separate the 'Smooth' from the 'Rough' with minimal use of formal mathematics or statistical methods, although these may be used on occasions. See also **initial data analysis**.

Exploratory factor analysis See **factor analysis**.

Exponential distribution A probability distribution of intervals between consecutive random events, i.e. those following a Poisson process. A simple example would be the time intervals between people arriving at a supermarket check-out queue. Also known as the *negative exponential distribution*.

> *Mathematical details*
>
> $$f(x) = \lambda \mathrm{e}^{-\lambda x}, \qquad x > 0$$
>
> The mean of the distribution is $1/\lambda$ and its variance is $1/\lambda^2$. Some examples of the distribution for different values of λ are shown in Figure 23.

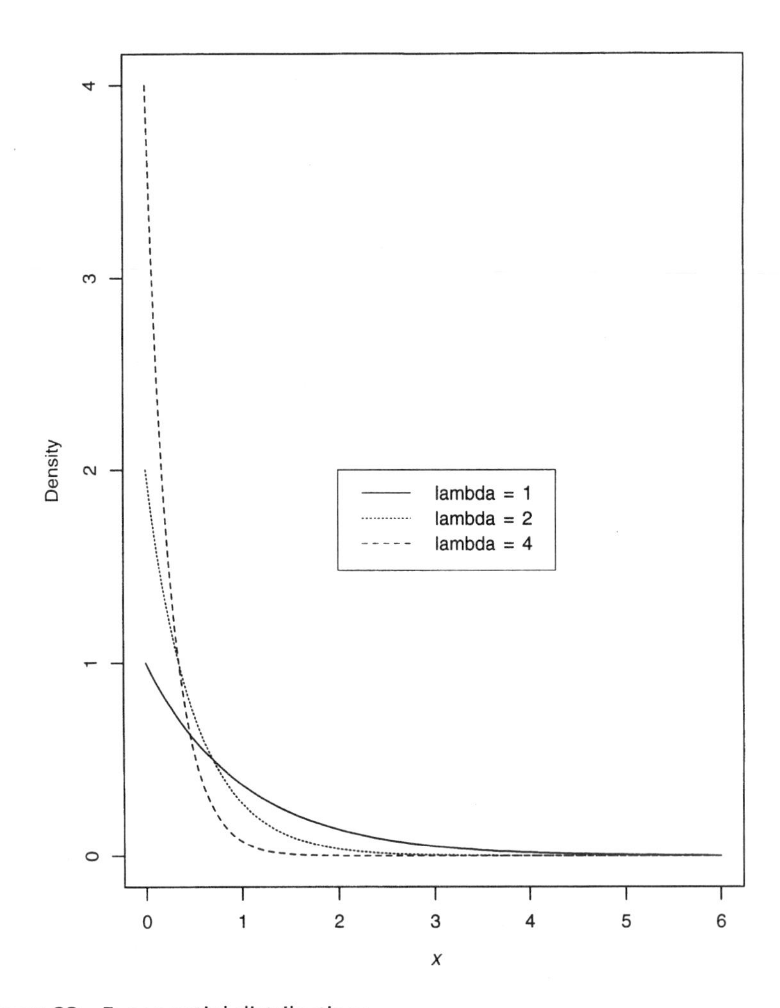

Figure 23 Exponential distributions.

Extrapolation The process of estimating from a data set those values lying beyond the range of the data. In **regression analysis**, for example, a value of the response variable may be estimated from the fitted equation for a new observation having values of the explanatory variables beyond the range of those used in deriving the equation. For example, predicting the weight of teenagers from their height, using a regression equation of weight on height developed from a sample of 8–10 year olds. In general a dangerous procedure.

Extreme values The largest and smallest variate values amongst a sample of observations. For example, in the sample of IQ values 95, 110, 112, 97, 89, 130, 150, the extreme values are 89 and 150.

Eyeball test Informal assessment of data simply by inspection and mental calculation allied with experience of the particular area from which the data arise.

F

Facets See **generalizability theory**.

Factor A term used in a variety of ways in statistics, but most commonly to refer to a **categorical variable**, with a small number of levels, under investigation in an experiment as a possible source of variation. Essentially, simply a categorical explanatory variable. Also used in **factor analysis** for a **latent variable** considered to be responsible for generating the correlations between the **manifest variables**.

Factor analysis A class of procedures for investigating the structure of the correlations or covariances between the variables in a set of **multivariate data**. Essentially the aim is to discover whether the correlations arise from the relationships of the observed or *manifest variables* to a small number of underlying, unobservable, *latent variables*, usually known in this context as the *common factors*. A number of approaches are used to estimate the parameters in the model (see below) and, after the initial estimation phase, an attempt is generally made to simplify the often difficult task of interpreting the derived factors using a process known as **factor rotation**. In general the aim is to provide a solution having what is known as *simple structure*, i.e. each common factor affects only a small number of the observed variables. Most often used in an exploratory fashion (*exploratory factor analysis*), but can be used to test specific hypotheses about the structure of the observed covariances (see **confirmatory factor analysis**).

Mathematical details

Factor analysis postulates that the correlations or covariances between a set of observed variables $\boldsymbol{x}' = [x_1, x_2, \ldots, x_q]$ arise from their relationship to the common factors $\boldsymbol{f}' = [f_1, f_2, \ldots, f_k]$ where $k < q$. Explicitly the model used is

$$\begin{aligned} x_1 &= \lambda_{11} f_1 + \lambda_{12} f_2 + \cdots + \lambda_{1k} f_k + u_1 \\ x_2 &= \lambda_{21} f_1 + \lambda_{22} f_2 + \cdots + \lambda_{2k} f_k + u_2 \\ &\vdots \\ x_q &= \lambda_{q1} f_1 + \lambda_{q2} f_2 + \cdots + \lambda_{qk} f_k + u_q \end{aligned}$$

This can be written concisely as

$$\boldsymbol{x} = \boldsymbol{\Lambda} \boldsymbol{f} + \boldsymbol{u}$$

where

$$\Lambda = \begin{pmatrix} \lambda_{11} & \lambda_{12} & \cdots & \lambda_{1k} \\ \lambda_{21} & \lambda_{22} & \cdots & \lambda_{2k} \\ \vdots & \vdots & \vdots & \vdots \\ \lambda_{q1} & \lambda_{q2} & \cdots & \lambda_{qk} \end{pmatrix}$$

contains the **regression coefficients** (usually known in this context as *factor loadings*) of the observed variables on the common factors, and $\boldsymbol{u}' = [u_1, u_2, \ldots, u_q]$ contains the specific variates and $\boldsymbol{f}' = [f_1, \ldots, f_k]$. The matrix Λ is known as the *loading matrix*. The model implies that the **variance–covariance matrix**, Σ, of the manifest variables is of the form

$$\Sigma = \Lambda \Lambda' + \Psi$$

where Ψ is a **diagonal matrix** containing the variances of the specific variates. A number of approaches are used to estimate the parameters in the model, i.e. the elements of Λ and Ψ, including **maximum likelihood estimation** assuming that the variables have a **multivariate normal distribution**.

Numerical example

The matrix below shows the correlations between nine statements about pain by 123 people suffering from extreme pain. Each statement was scored on a scale of 1 to 6 ranging from disagreement to agreement:

	Statement								
	1	2	3	4	5	6	7	8	9
1	1.00								
2	−0.04	1.00							
3	0.61	−0.07	1.00						
4	0.45	−0.12	0.59	1.00					
5	0.03	0.49	0.03	−0.08	1.00				
6	−0.29	0.43	−0.13	−0.21	0.47	1.00			
7	−0.30	0.30	−0.24	−0.19	0.41	0.63	1.00		
8	0.45	−0.31	0.59	0.63	−0.14	−0.13	−0.26	1.00	
9	0.30	−0.17	0.32	0.37	−0.24	−0.15	−0.29	0.40	1.00

The nine statements were as follows:

1. Whether or not I am in pain in the future depends on the skills of the doctors.
2. Whenever I am in pain, it is usually because of something I have done or not done.
3. Whether or not I am in pain depends on what the doctors do for me.
4. I cannot get any help for my pain unless I go to seek medical advice.
5. When I am in pain I know that it is because I have not been taking proper exercise or eating the right food.
6. People's pain results from their own carelessness.
7. I am directly responsible for my pain.
8. Relief from pain is chiefly controlled by the doctors.
9. People who are never in pain are just plain lucky.

The two-factor solution produced by maximum likelihood factor analysis is:

	Estimated loadings			
Statement	Factor 1	Factor 2	Communality	Specific variance
1	0.64	0.21	0.46	0.54
2	−0.36	0.41	0.30	0.70
3	0.72	0.40	0.68	0.32
4	0.69	0.29	0.56	0.44
5	−0.31	0.57	0.42	0.58
6	−0.52	0.61	0.45	0.35
7	−0.56	0.48	0.54	0.46
8	0.71	0.25	0.57	0.43
9	0.48	0.00	0.23	0.77
Variance	2.95	1.45		

The second factor is positively correlated with all nine statements – might be labelled general pain level. The first factor is negatively correlated with statements taking personal responsibility for one's pain and positively correlated with statements in which the control of, and reasons for, pain are attributed elsewhere. This is a **bipolar factor** that might be easier to interpret after **factor rotation**.

See also **structural equation modelling** and **principal components analysis**. [*Psychological Assessment*, 1998, 10(2), 83–89].

Factorial designs Designs which allow two or more questions to be addressed in an investigation. The simplest factorial design is one in which each of two factors is either present or absent. Such designs enable possible **interactions** between factors to be investigated.

Mathematical details

Model assumed for the observations y_{ijk} in a two factor design:

$$y_{ijk} = \mu + \alpha_i + \beta_j + \gamma_{ij} + \varepsilon_{ijk}$$

where μ is overall mean, α_i is factor 1 effect, β_j is factor 2 effect, γ_{ij} is the interaction effect and ε_{ijk} are error terms having a **normal distribution** with mean zero and variance σ^2.

Numerical example

The data below arise from an investigation into types of slimming regime. In this case the two factor variables are: treatment – with two levels, namely whether or not a woman was advised to use a slimming manual based on psychological behaviourist theory as an addition to the regular package offered by the clinic – and status, also with two levels: 'novice' and 'experienced' slimmer. The dependent variable here is a measure of weight change over 3 months, with negative values indicating a decrease in weight.

Slimming data

	Novice	Experienced
No manual	−2.85	−2.42
	−1.98	0.00
	−2.12	−2.74
	0.00	−0.84
Manual	−4.44	0.00
	−8.11	−1.64
	−9.40	−2.40
	−3.50	−2.15

Analysis of variance table

Source	SS	df	MS	*F*	*P*
Condition	21.83	1	21.83	7.04	0.021
Status	25.53	1	25.53	8.24	0.014
Condition × status	20.95	1	20.95	6.76	0.023
Error	37.19	12	3.10		

The significant interaction reflects the fact that the decrease in weight produced by giving novice slimmers access to the slimming manual is far greater than that achieved with experienced slimmers.

Factor loading See **factor analysis**.

Factor rotation The final stage of a **factor analysis**, in which the factors derived initially are transformed to make their interpretation simpler. In general, the aim of the process is to make the common factors more clearly defined, by increasing the size of large factor loadings and decreasing the size of those that are small. **Bipolar factors** are generally split into two separate parts, one corresponding to those variables with positive loadings and the other to those variables with negative loadings. The most commonly used method is **varimax rotation** which seeks to optimize a function of the variance of the squared factor loadings.

Numerical example

In a small factor analysis example with four variables the initial two-factor solution was as follows:

	F1	F2
Variable 1	−0.40	0.90
Variable 2	0.25	−0.95
Variable 3	0.93	0.35
Variable 4	0.96	0.29

Applying varimax rotation leads to the following rotated solution:

	F1	F2
Variable 1	−0.09	−0.98
Variable 2	−0.07	0.98
Variable 3	0.99	−0.03
Variable 4	0.99	0.04

[*Educational and Psychological Measurement*, 1996, 56, 460–474].

Fair game A game in which the entry cost or stake equals the expected gain. In a sequence of such games between two opponents the one with the larger capital has the greater chance of ruining his opponent.

Familywise error rate The probability of making any error in a given family of inferences. See also **pre-comparison error rate** and **per-experiment error rate**.

***F*-distribution** The **probability distribution** of the ratio of two independent random variables, each having a **chi-squared distribution**, divided by their respective degrees of freedom. Of fundamental importance in **analysis of variance** where it is the distribution of mean square terms under the hypothesis of equality of means.

Feasibility study Essentially a synonym for **pilot study**.

Finite mixture distribution A **probability distribution** constructed as a weighted linear function of a number of component probability distributions. Such distributions are used to model populations thought to contain relatively distinct groups. For example, in a sample of depressed patients thought to consist of psychotic and neurotic groups, a measure of depression might be modelled by the following mixture of two normal distributions:

$$f(\text{depression}) = pN(\mu_1, \sigma_1) + (1 - p)N(\mu_2, \sigma_2)$$

where p is the proportion of psychotic depressed patients in the population, having depression scores with mean μ_1 and standard deviation σ_1, and $1 - p$ is the proportion of neurotic depressed patients, having depression scores with mean μ_2 and standard deviation σ_2. The parameters in the model would usually be estimated by applying **maximum likelihood estimation** to a sample of the depression scores. Mixtures of **multivariate normal distributions** are often used as the basis of **cluster analysis**. Figure 24 shows an example of a finite mixture distribution with two normal components.

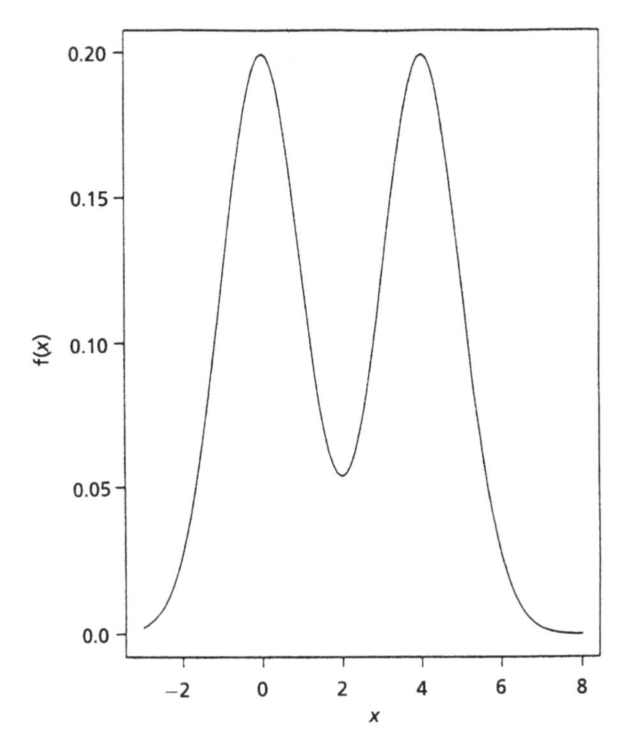

Figure 24 An example of a finite mixture distribution; a distribution with two normal distributions as components.

Finite population A population of finite size. For example the number of professors of pyschology in the UK.

Fisher's exact test An alternative method to the **chi-squared statistic** for assessing the independence of two variables forming a **two-by-two contingency table**, particularly when the **expected frequencies** are small. The method consists of evaluating the sum of the probabilities associated with the observed table and all possible two-by-two tables that have the same row and column totals as the observed data, but exhibit more extreme departure from independence. The probability of each table is calculated from the **hypergeometric distribution**.

Numerical example

The table below shows a count of the number of patients with suicidal feelings amongst two groups of depressed patients (psychotic and neurotic depression):

	Psychotic	Neurotic	Total
Suicidal feelings	2	6	8
No suicidal feelings	18	14	32
Total	20	20	40

The probability of the observed table is calculated as

$$P^{(1)} = \frac{8! \times 32! \times 20! \times 20!}{2! \times 6! \times 18! \times 14! \times 20!} = 0.095760$$

The probabilities of the two tables representing more extreme departures from independence are

$$P^{(2)} = 0.020160 \quad \text{and} \quad P^{(3)} = 0.001638$$

Consequently the probability of obtaining the observed arrangement or one more suggestive of a departure from independence is 0.11756. The data give no evidence that psychotics and neurotics differ with respect to having suicidal feelings.

Fisher's linear discriminant function A technique for deriving a rule with which to classify an individual into one of two possible groups on the basis of a set of measurements made on the individual. The essence of the approach is to seek a linear transformation of the measurements such that the ratio of the between group variance to the within group variance of the transformed variable is maximized.

Mathematical details

A linear transformation, $z = a_1x_1 + a_2x_2 + \cdots + a_qx_q$, of the observed variables $x_1, x_2, \ldots, x_q$ is sought such that the separation between the group means on the transformed scale, $\bar{z}_1$ and $\bar{z}_2$, is maximized relative to the within group variation on the z scale. The solution can be shown to be

$$\boldsymbol{a} = \mathbf{S}^{-1}(\bar{\boldsymbol{x}}_1 - \bar{\boldsymbol{x}}_2)$$

where $\mathbf{S}$ is the pooled within group variance—covariance matrix of the two groups and $\bar{\boldsymbol{x}}_1$, $\bar{\boldsymbol{x}}_2$ are the group mean vectors.

Numerical example

Below are given the means and pooled covariance matrix for three psychological tests administered to patients suffering from anxiety and depression.

Means

	Test			
	1	**2**	**3**	***N***
Anxiety	2.93	1.67	0.73	100
Depression	3.03	1.24	0.54	100

Covariance matrix

$$\mathbf{S} = \begin{bmatrix} 2.30 & & \\ 0.25 & 0.61 & \\ 0.47 & 0.04 & 0.60 \end{bmatrix}$$

The coefficients of the discriminant function are

$$-0.22, 0.76, 0.43$$

Consequently the discriminant mean for the anxiety group is 0.94 and for the hysteria group is 0.51. The classification rule for assigning a new patient becomes: assign to anxiety group if $-0.22\,x_1 + 0.76\,x_2 + 0.43\,x_3 > 0.725$, otherwise assign to the depression group.

Fisher's *z* transformation A transformation of **Pearson's product moment correlation** coefficient r designed to achieve normality and to simplify testing hypotheses about the population correlation coefficient.

Mathematical details

$$z = \frac{1}{2}\ln\frac{1+r}{1-r}$$

The statistic z has mean $\frac{1}{2}\ln\frac{1+\rho}{1-\rho}$, where ρ is the population correlation value, and variance $\frac{1}{n-3}$, where n is the sample size. The transformation may be used to test hypotheses about ρ and to construct associated **confidence intervals**.

Numerical example

Suppose for a sample of 40 individuals from the same population the correlation between reaction time and IQ is calculated to be 0.7. Then we have

$$z = \frac{1}{2}\ln\frac{1+0.7}{1-0.7} = 0.867$$

A 95% confidence interval for the mean of z, i.e. $\frac{1}{2}\ln\frac{1+\rho}{1-\rho}$ can be obtained simply in the usual way as $z \pm 1.96\,\mathrm{sd}(z)$, i.e. $z \pm 1.96\frac{1}{\sqrt{n-3}}$. In the example this leads to the interval $(0.55, 1.19)$. This can be changed into the more useful confidence interval for ρ, the population correlation coefficient, itself by rewriting the original transformation in an alternative fashion, namely

$$\rho = \frac{e^{2z}-1}{e^{2z}+1}$$

The confidence interval for ρ is now obtained by simply substituting the previously derived confidence interval limits to give

$$\frac{e^{2\times 0.5451}-1}{e^{2\times 0.5451}+1}, \frac{e^{2\times 1.1895}-1}{e^{2\times 1.1895}+1} = (0.50,\ 0.83)$$

Fishing expedition Synonym for **data dredging**.

Five-number summary A method of summarizing a set of observations using the minimum value, the lower quartile, the median, upper quartile and maximum value. Forms the basis of the box-and-whisker plot.

Numerical example

Suppose the following IQ scores were obtained on a sample of 20 first year psychology students:

98, 101, 110, 99, 120, 105, 106, 106, 95, 115, 101, 104, 125, 110, 93, 98, 107, 110, 109, 102

Then the five-number summary is:

- Minimum = 93
- Lower quartile = 100
- Median = 105.5
- Upper quartile = 110
- Maximum = 125

Fixed effects The effects attributable to a finite set of levels of a factor that are of specific interest. For example, the investigator may wish to compare the effects of three particular diets on reading speed. *Fixed-effects models* are those that contain only factors with this type of effect. See also **random effects**.

Fixed-effects model See **fixed effects**.

Floor effect See **ceiling effect**.

Follow-up The process of locating participants in a study after the study has been completed, to determine whether or not some outcome of interest has occurred.

Forward-looking study Synonym for **prospective study**.

Forward selection See **selection methods in regression**.

Fourfold table Synonym for **two-by-two contingency table**.

Freeman–Tukey transformation A transformation of a random variable, x, having a Poisson distribution, to the form $\sqrt{x} + \sqrt{x+1}$ in order to stabilize its variance.

Frequency distribution The division of a sample of observations into a number of classes, together with the number of observations in each class. Acts as a useful summary of the main features of the data, such as location, shape and spread. An example of such a table is as follows:

Social functioning scores	
Class limits	Observed frequency
75–79	1
80–84	2
85–89	5
90–94	9
95–99	10
100–104	7
105–109	4
110–114	2
≥115	1

See also **histogram**.

Frequentist inference An approach to statistics based on a frequency view of probability in which it is assumed that it is possible to consider an infinite sequence of independent repetitions of the same statistical experiment. Significance tests, hypothesis tests and **likelihood** are the main tools of this form of inference. See also **Bayesian inference**.

Friedman's two-way analysis of variance A **distribution-free method** that is the analogue of the **analysis of variance** for a design with two factors. Can be applied to data sets that do not meet the assumptions of the parametric approach: namely, normality and homogeneity of variance. Uses only the ranks of the observations. [*Neuropsychobiology*, 1997, 35, 46–50].

***F*-test** A test for the equality of the variances of two populations having **normal distributions**, based on the ratio of the variances of a sample of observations taken from each. Most often encountered in the **analysis of variance**, where testing whether particular variances are the same, tests indirectly for the equality of a set of means. The *F*-test is also often encountered prior to the application of **Student's *t*-test**, to assess the assumption that the two populations have the same variance.

***F*-to-enter** See **selection methods in regression**.

***F*-to-remove** See **selection methods in regression**.

Funnel plot An information method of assessing the effect of **publication bias**, usually in the context of a **meta-analysis**. The effect measures from each reported study are plotted on the *x*-axis against, on the *y*-axis, the corresponding sample sizes. Because of the nature of sampling variability, this plot should, in the absence of publication bias, have the shape of a pyramid with a tapering 'funnel-like' peak. Publication bias will tend to skew the pyramid by selectively excluding studies with small or no significant effects. Such studies predominate when the sample sizes are small, but are increasingly less common as the sample sizes increase. Therefore, their absence removes part of the lower left-hand corner of the pyramid. This effect is illustrated in the plots shown in Figure 25. [*Psychological Methods*, 1998, 3, 46–54].

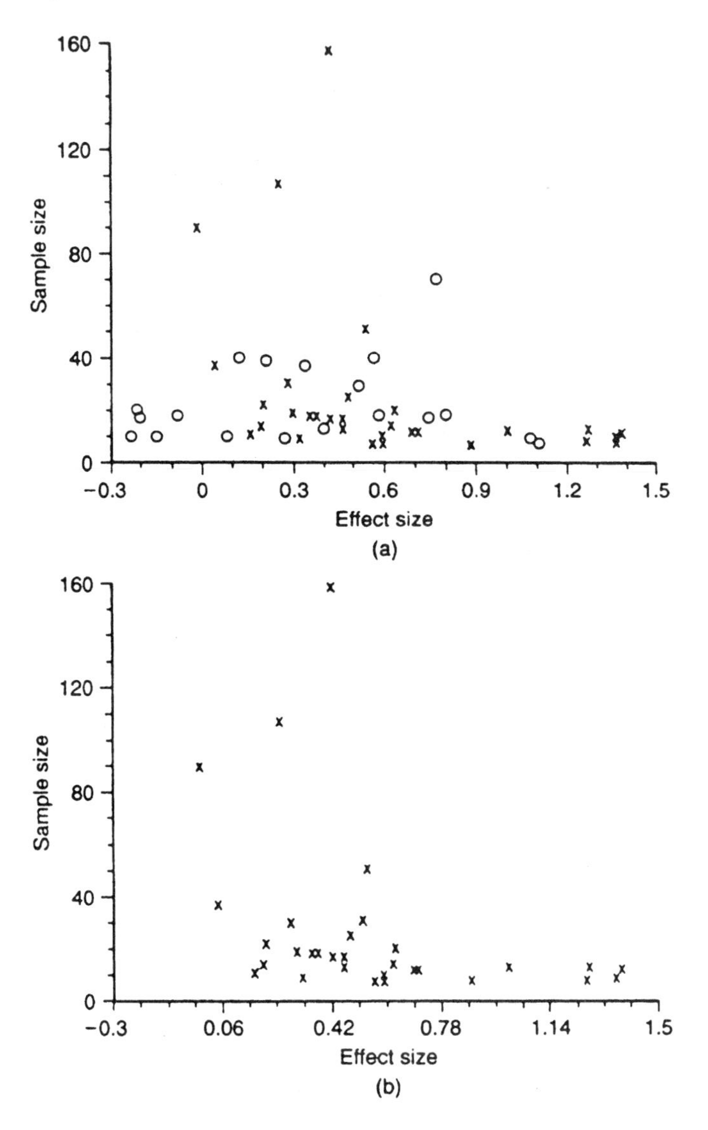

Figure 25 Funnel plot of studies of psychoeducational progress for surgical patients: (a) all studies; (b) published studies only.

G

Gambler's fallacy The belief that if an event has not happened for a long time, it is bound to occur soon.

Game theory The branch of mathematics that deals with the theory of contests between two or more players under specified sets of rules. The subject assumes a statistical aspect when part of the game proceeds under a chance scheme.

Gamma distribution A **probability distribution** that is often used for modelling data having positive **skewness**.

Mathematical details

The formula defining the distribution is:

$$f(x) = \frac{x^{\alpha-1} e^{-x}}{\Gamma(\alpha)}, \qquad x > 0$$

where Γ is the **gamma function**. The mean and variance of the distribution are both equal to the parameter α. Some examples of gamma distribution for various values of α are shown in Figure 26.

Gamma function The function Γ defined by

$$\Gamma(r) = \int_0^\infty t^{r-1} e^{-t}\, dt$$

where $r > 0$ (r need not be an integer). The function is recursive, satisfying the relationship $\Gamma(r+1) = r\Gamma(r)$. For integer r, $\Gamma(r+1) = r!$. Used in defining **gamma distribution**.

Garbage in, garbage out A term that draws attention to the fact that sensible output only follows from sensible input. Specifically, if the data are originally of dubious quality then so also will be the results.

Gaussian distribution Synonym for **normal distribution**.

Generalizability theory A theory of measurement that recognizes that in any measurement situation there are multiple (in fact infinite) sources of variation (called *facets* in the theory), and that an important goal of measurement is to attempt to identify and measure **variance components** which are contributing error to an estimate. Strategies can then be implemented to reduce the influence of these sources on the measurement. [*Journal of Experimental Education*, 1997, 65, 367–379].

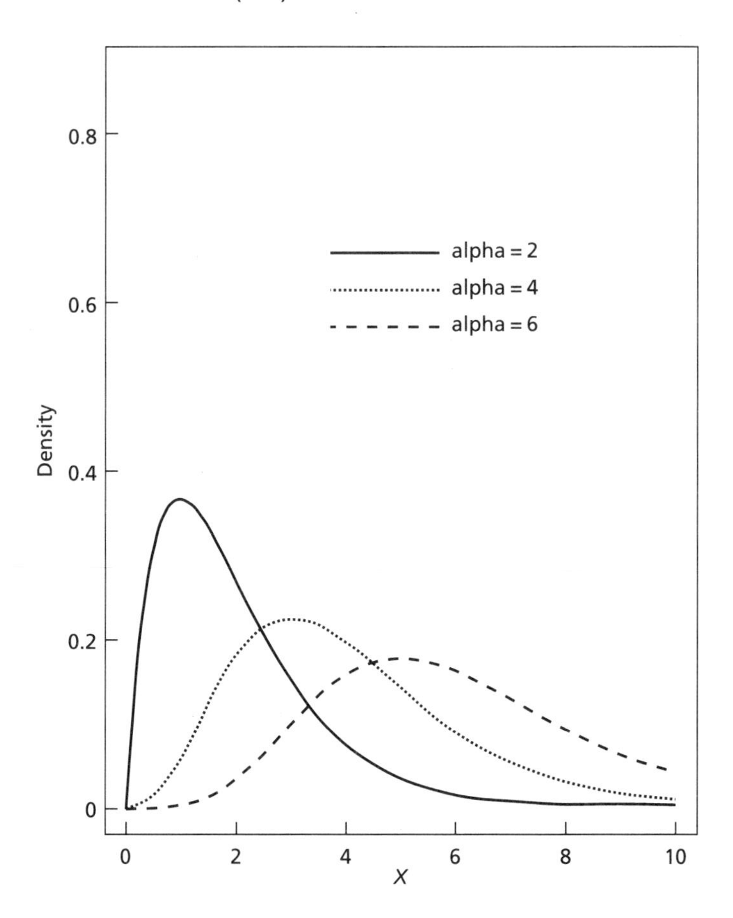

Figure 26 Gamma distributions.

Generalized linear models (GLM) A class of models that arise from a generalization of ordinary linear regression. Instead of the expected value of a dependent variable being modelled as a linear function of explanatory variables, some function (known as the *link function*) of the expected value is modelled. So, for example, it might, for some types of response variables, be appropriate to postulate that the logarithm of the expected value of the dependent variable was a linear function of the explanatory variables. In addition, other distributions than the normal are allowed for the response variable. Includes multiple regression, logistic regression and Poisson regression as special cases.

Generalized variance An analogue of the variance for summarizing the spread of a set of multivariate data. Given by the determinant of the variance–covariance matrix of the observations.

Numerical example

(a) For a bivariate data set with variance–covariance matrix given by

$$\mathbf{S} = \begin{bmatrix} 1.0 & 0.0 \\ 0.0 & 1.0 \end{bmatrix}$$

the generalized variance is $\det(\mathbf{S}) = 1 \times 1 - 0 \times 0 = 1$.

(b) For a bivariate data set with variance–covariance matrix given by

$$\mathbf{S} = \begin{bmatrix} 1.0 & 0.8 \\ 0.8 & 1.0 \end{bmatrix}$$

the generalized variance is $\det(\mathbf{S}) = 1 \times 1 - 0.8 \times 0.8 = 0.36$.

GENSTAT A general-purpose piece of statistical software for the management and analysis of data. The package incorporates a wide variety of data-handling procedures and a wide range of statistical techniques, including **multiple regression**, **cluster analysis**, and **principal components analysis**. It uses a sophisticated statistical programming language that enables non-standard methods of analysis to be implemented relatively easily.

Geometric distribution The **probability distribution** of the number of trials R in a series of independent trials each of which can result in one of two alternatives (S or F say), until the first S occurs.

Mathematical details

The formula for the distribution is

$$\Pr(R = r) = p(1 - p)^{r-1}, \qquad r = 1, 2, \ldots$$

where $p = \Pr(S)$ and Pr denotes probability. The mean of the distribution is $1/p$ and the variance is $(1 - p)/p^2$.

Numerical example

In tossing a coin, the probability that there are 4 tails before the first head is

$$\Pr(r = 5) = \tfrac{1}{2} \times (\tfrac{1}{2})^4 = \tfrac{1}{32} = 0.031$$

Geometric mean A measure of location that is useful as an alternative to the mean as a measure of location when observed values are thought to be the results of many multiplicative (rather than additive) random effects.

Mathematical details

Given observations $x_1, x_2, \ldots, x_n$, the geometric mean g is calculated as

$$g = \left(\prod_{j=1}^{n} x_j \right)^{1/n}$$

Numerical example

For the set of 10 observations 6, 4, 8, 3, 7, 6, 5, 8, 10, 1, the geometric mean is given by

$$g = [6 \times 4 \times 8 \times 3 \times 7 \times 6 \times 5 \times 8 \times 10 \times 1]^{1/10} = 4.995$$

The arithmetic mean in contrast is 5.800.

GLIM A software package particularly suited for fitting **generalized linear models** (the acronym stands for Generalized Linear Interactive Modelling), including **log-linear models** and **logistic regression**. A large number of GLIM macros are now available that can be used for a variety of non-standard statistical analyses.

GLM Abbreviation for **generalized linear model**.

Glyphs A graphical representation of **multivariate data**, in which each observation is represented by a circle, with rays of different lengths indicating the values of the observed variables. An example is given in Figure 27. See also **Chernoff's faces**.

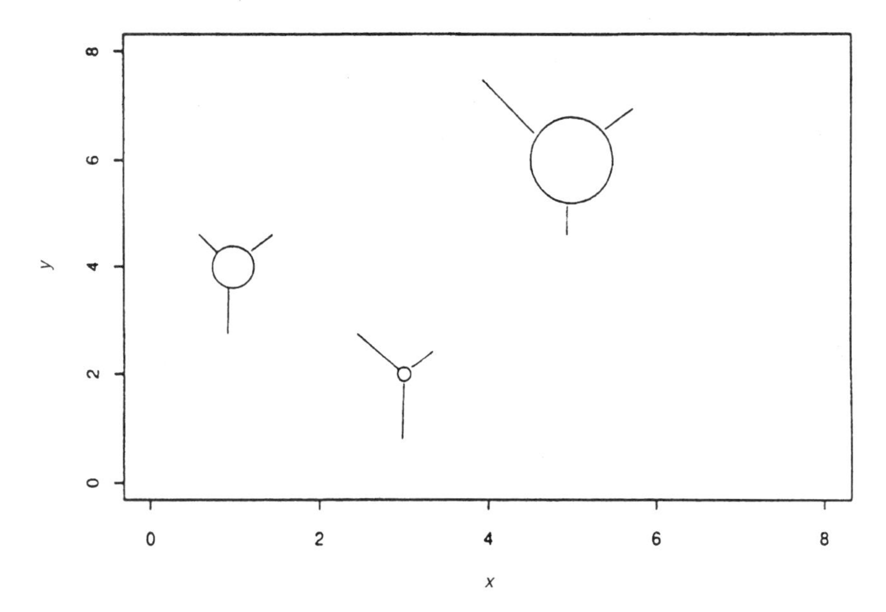

Figure 27 Glyphs.

Goodman and Kruskal measures of association Measures of associations that are useful in the situation where two categorical variables cannot be assumed to be derived from perhaps unobservable continuous variables and where there is no natural ordering of interest. The rationale behind the measures is the question: 'How much does knowledge of the classification of one of the variables improve the ability to predict the classification of the other variable?'

Goodness-of-fit statistics Measures of the agreement between a set of sample observations and the corresponding values predicted from some model of interest. Many such measures have been suggested. See **chi-squared statistic**, **deviance** and **likelihood ratio**.

Grand mean Mean of all the values in a grouped data set irrespective of groups.

Graphical methods A generic term for those techniques in which the results are given in the form of a graph, diagram or some other form of visual display. Examples are **Chernoff's faces** and **glyphs**.

Greenhouse–Geisser correction A method of adjusting the degrees of freedom of the within subject **F-tests** in the **analysis of variance** of **longitudinal data**, so as to allow for possible departures of the **variance–covariance matrix** of the measurements from the assumption of **sphericity**. If this condition holds for the data, then the correction factor is unity and the simple *F*-tests are valid. Departures from sphericity result in an estimated correction factor less than unity, thus reducing the degrees of freedom of the relevant *F*-tests. Now primarily of historical interest since there are more suitable approaches to the analysis of longitudinal data, for example **multilevel models**. See also **Huynh–Feldt correction**, **compound symmetry** and **Mauchly test**. [*Journal of Clinical Psychology*, 1996, 52, 243–252].

Grouped data Data recorded as frequencies of observations in particular intervals.

H

H_0 Symbol for null hypothesis.

H_1 Symbol for alternative hypothesis.

Halo effect The tendency of a subject's performance on some task to be overrated because of the observer's perception of the subject's 'doing well' gained in an earlier exercise or when assessed in a different area. [*Journal of Applied Psychology*, 1997, 82, 665–674].

Hanging rootogram A diagram comparing an observed **rootogram** with a fitted curve, in which differences between the two are displayed in relation to the horizontal axis, rather than to the curve itself. This makes it easier to spot large differences and to look for patterns. An example is shown in Figure 28.

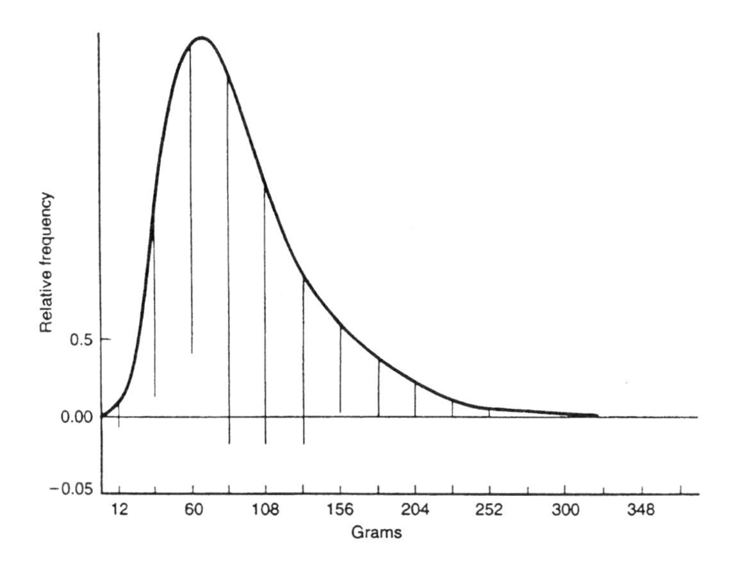

Figure 28 Hanging rootogram.

Hardy–Weinberg law The law stating that both gene frequencies and genotype frequencies will remain constant from generation to generation in an infinitely large interbreeding population in which mating is at random and there is no selection, migration or mutation. In a situation where a single pair of alleles (A and a) is considered, the frequencies of germ cells carrying A and a are defined as p and q, respectively. At equilibrium the frequencies of the genotype classes are $p^2(AA)$, $2pq(Aa)$ and $q^2(aa)$.

Harmonic mean A statistic, H, which is the reciprocal of the arithmetic mean of the reciprocals of a set of observations $x_1, x_2, \ldots, x_n$. Specifically obtained from the equation

$$\frac{1}{H} = \frac{1}{n} \sum_{i=1}^{n} \frac{1}{x_1}$$

Used in some methods of analysing **unbalanced design**. The harmonic mean is either smaller than or equal to the arithmetic mean.

Numerical example

For the set of 10 observations 6, 4, 8, 3, 7, 6, 5, 8, 10, 1, the harmonic mean is given by

$$\frac{1}{H} = \frac{1}{10}\left[\frac{1}{6} + \frac{1}{4} + \frac{1}{8} + \frac{1}{3} + \frac{1}{7} + \frac{1}{6} + \frac{1}{5} + \frac{1}{8} + \frac{1}{10} + \frac{1}{1}\right] = 0.261$$

Therefore $H = 3.832$. The arithmetic mean of the observation is 5.800.

Hartley's test A simple test of the equality of variances of the populations corresponding to the groups in a **one-way design**.

Mathematical details

The **test statistic** (if each group has the same number of observations) is the ratio of the largest (s^2 largest) to the smallest (s^2 smallest) within group variance, i.e.

$$F = \frac{s^2 \text{ largest}}{s^2 \text{ smallest}}$$

Critical values are available in many statistical tables. The test is sensitive to departures from normality.

Numerical example

In the teaching methods example under the **analysis of variance** entry the variances of the four groups are:

Method 1	Method 2	Method 3	Method 4
12.75	6.03	9.53	13.11

The test statistic for Hartley's test is therefore:

$$F_{\text{max}} = 13.11/6.03 = 2.17$$

The critical value of the test statistic at the 5% significance level is 7.18. Consequently, there is no evidence of a difference in the variances of the four groups.

See also **Bartlett's test**, **Box's test** and **Levine's test**.

Hat matrix A matrix encountered in **multiple regression**, which, applied to the observed values, gives the corresponding predicted values.

> *Mathematical details*
>
> The hat matrix, $\mathbf{H}$, is found from the **design matrix**, $\mathbf{X}$, as
>
> $$\mathbf{H} = \mathbf{X}(\mathbf{X}'\mathbf{X})^{-1}\mathbf{X}$$
>
> The predicted values $\hat{y}$ are found from the observed values as $\hat{y} = \mathbf{H}y$. Arises in calculating various types of **residuals**.

Hawthorne effect A term used for the effect that might be produced in an experiment simply from the awareness by the subjects that they are participating in some form of scientific investigation. The name comes from a study of industrial efficiency at the Hawthorne Plant in Chicago in the 1920s. [*Journal of Computing in Childhood Education*, 1994, 5, 61–71].

Hazard function A function used in the analysis of **survival times**, where it is often of interest to assess which periods have the highest and lowest chance of death (or whatever the event of interest happens to be) amongst those alive at the time. In the very old, for example, there is a high risk of dying each year, amongst those entering that stage of their life. The probability of any individual dying in their 100th year is, however, small because so few individuals live to be 100 years old. The function, which is usually denoted $h(t)$, is defined as the risk that an individual experiences an event (death, improvement, etc.) in a small time interval, given that the individual has survived up to the beginning of the interval. It is a measure of how likely an individual is to experience an event, as a function of the age of the individual. The hazard function may remain constant, increase, decrease, or take on some more complex shape. The hazard function for death in human beings, for example, has the 'bath tub' shape shown in Figure 29. It is relatively high immediately after birth, declines rapidly in the early years and then remains approximately constant before beginning to rise again during late middle age. The function can be estimated as the proportion of individuals

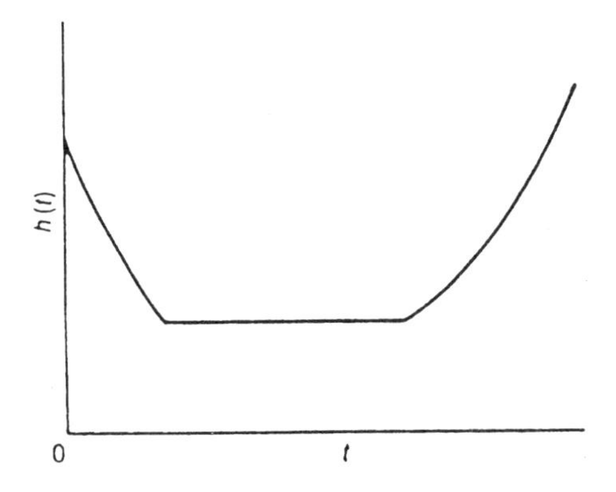

Figure 29 Bath tub shaped hazard function.

experiencing an event in an interval per unit time, given that they have survived to the beginning of the interval; that is,

$$\hat{h}(t) = \frac{\text{number of individuals experiencing an event in interval beginning at } t}{(\text{number of individuals surviving at } t)(\text{interval width})}$$

Care is needed in the interpretation of the hazard function, because of selection effects due to both variation between individuals and variation within each individual over time. For example, individuals with a high risk are more prone to experience an event early, and those remaining at risk will tend to be a selected group with a lower risk. This will result in the hazard rate being 'pulled down' to an increasing extent as time passes. See also **survival function**. [*Issues in Criminological and Legal Psychology*, 1995, 22, 77–83].

Healthy worker effect The phenomenon whereby employed individuals tend to have lower mortality rates than those unemployed. The effect, which can pose a serious problem in the interpretation of industrial **cohort studies**, has two main components:

- selection at recruitment to exclude the chronically sick, resulting in low standardized mortality rates among recent recruits to an industry;
- a secondary selection process, by which workers who become unfit during employment tend to leave, again leading to lower standardization mortality ratios amongst long-serving employees.

Helmert contrast A **contrast** often used in the **analysis of variance**, in which each level of a factor is tested against the average of the remaining levels. So, for example, if three groups are involved, of which the first is a control and the other two are treatment groups, the first contrast tests the control group against the average of the two treatments, and the second tests whether the two treatments differ.

Heterogeneous A term used in statistics to indicate the inequality of some quantity of interest (usually a variance) in a number of different groups, populations, etc. See also **homogeneous**.

Heuristic computer program A computer program which attempts to use the same sort of selectivity in searching for solutions that human beings use.

Heywood cases Solutions obtained when using **factor analysis** in which one or more of the variances of the **specific variates** become negative.

Hierarchical clustering See **agglomerative hierarchical clustering**.

Hierarchical models A series of models for a set of observations, where each model results from adding or deleting parameters from other models in the series. For example, consider models for a two-way design with factors A and B; the following models are hierarchical:

$$A + B + AB$$
$$A + B$$
$$A$$

Higgin's law A 'law' that states that the prevalence of any condition is inversely proportional to the number of experts whose agreement is required to establish its presence.

Hill-climbing algorithm An algorithm used in those techniques of **cluster analysis** which seek to find a partition of n individuals into g clusters by optimizing some numerical index of clustering. For example, for **univariate data**, choosing the partition that minimizes the within group variance might be sought. Since it is impossible to consider every partition of the n individuals into g groups (because of the enormous number of partitions), the algorithm begins with some given initial partition and considers individuals in turn for moving into other clusters, making the move if it causes an improvement in the value of the clustering index. The process is continued until no move of a single individual causes an improvement.

Hinge A more exotic (but less desirable) term for **quartile**.

Histogram A graphical representation of a set of observations, in which class frequencies are represented by the areas of rectangles centred on the class interval. If the latter are all equal, the heights of the rectangles are also proportional to the observed frequencies. An example of a histogram, involving the IQ scores of 200 psychology students, is given in Figure 30.

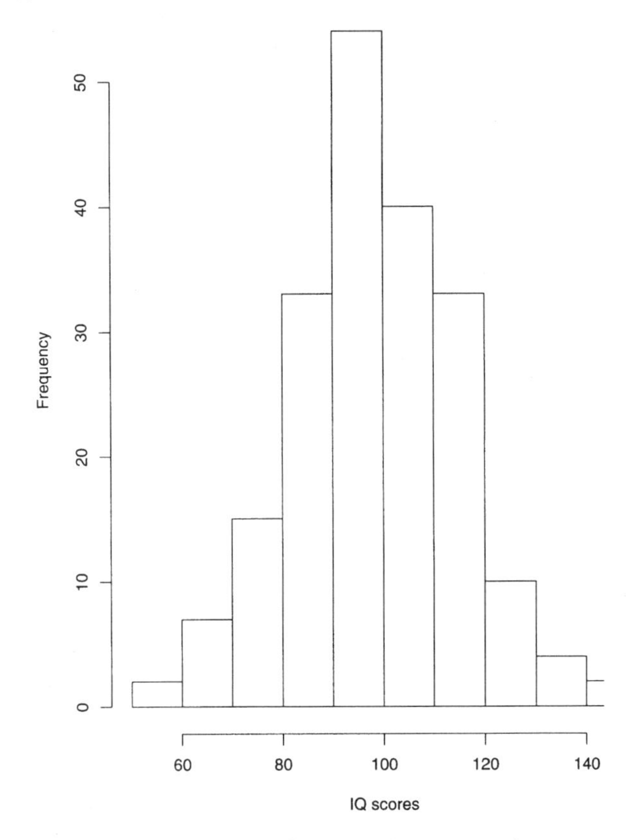

Figure 30 Histogram of IQ scores of 200 psychology students.

Historical controls A group of patients treated in the past with a standard therapy, used as the control group for evaluating a new treatment on current patients. Although used fairly frequently in psychological investigations, the approach is not to be recommended since possible **biases**, due to other factors that may have changed over time, can never be satisfactorily eliminated. See also **literature controls**.

Hit rate A term occasionally used for the number of correct classifications in a **discriminant analysis**. An optimistic estimate of the true correct classification rate. See also **jackknife**.

Holdover effect Synonym for **carryover effect**.

Homogeneous A term that is used in statistics to indicate the equality of some quantity of interest (most often a variance), in a number of different groups, populations, etc. See also **heterogeneous**.

Hospital controls See **control group**.

Hot deck A method widely used in surveys for the **imputation** of **missing values**. In its simplest form the method involves sampling with replacement m values from the sample respondents A_r to an item y when m is the number of non-respondents and r is the number of respondents. The sampled values are used in place of the missing values. In practice, the accuracy of imputation is improved by forming two or more imputation classes using control variables observed in all sample units, and then applying the procedure separately within each imputation class for each item with missing values.

Hotelling's T^2 test A generalization of **Student's t-test** for **multivariate data**. Can be used to test whether the population **mean vector** of a set of q variables is the **null vector**, or whether the mean vectors of two populations are equal.

Mathematical details

To test the hypothesis

$$H_0: \ \boldsymbol{\mu}_1 = \boldsymbol{\mu}_2$$

where $\boldsymbol{\mu}_1$ and $\boldsymbol{\mu}_2$ are the population mean vectors of two groups of interest, we use the test statistic

$$T^2 = \frac{n_1 n_2 (\bar{\mathbf{x}}_1 - \bar{\mathbf{x}}_2)' \mathbf{S}^{-1} (\bar{\mathbf{x}}_1 - \bar{\mathbf{x}}_2)}{n_1 + n_2}$$

where n_1 and n_2 are sample sizes, $\bar{\mathbf{x}}_1$ and $\bar{\mathbf{x}}_2$ are sample mean vectors, and $\mathbf{S}$ is a weighted average of the separate sample **variance–covariance matrices** and is given by

$$\mathbf{S} = \frac{(n_1 - 1)\mathbf{S}_1 + (n_2 - 1)\mathbf{S}_2}{n_1 + n_2 - 2}$$

where $\mathbf{S}_1$ and $\mathbf{S}_2$ are the covariance matrices of the two samples. Under the hypothesis that the population mean vectors are the same, and assuming the variables have a **multivariate normal distribution** and that $\mathbf{S}_1 = \mathbf{S}_2$, then

$$\frac{n_1 + n_2 - q - 1}{(n_1 + n_2 - 2)q} T^2$$

has an **F-distribution** with q and $(n_1 + n_2 - q - 1)$ degrees of freedom.

Numerical example

The data below give the results from four psychological tests on 10 men and 10 women:

	Test					Test			
Women	1	2	3	4	Men	1	2	3	4
1	15	17	24	14	1	13	14	12	21
2	17	15	32	26	2	14	12	14	26
3	15	14	29	23	3	12	19	21	21
4	13	12	10	16	4	12	13	10	16
5	20	17	26	28	5	11	20	16	16
6	15	21	26	21	6	12	9	14	18
7	15	13	26	22	7	10	13	18	24
8	13	5	22	22	8	10	8	13	23
9	14	7	30	17	9	12	20	19	23
10	17	15	30	27	10	11	10	11	27

The tests were as follows:

1. pictorial inconsistencies
2. paper form board
3. tool recognition
4. vocabulary

Hotelling's T^2 test takes the value 65.63 with an associated **P-value** 0.001. The data give strong evidence of a difference in the mean vectors of these four tests in men and women. The latter tend to have higher scores on all four tests.

[*Journal of Occupational Rehabilitation*, 1997, 7, 45–51].

Huynh–Feldt correction A correction term applied in the analysis of data from **longitudinal studies** by simple **analysis of variance** procedures, to ensure that the within subject **F-tests** are approximately valid even if the assumption of **sphericity** is invalid. See also **Greenhouse–Geisser correction**, **Mauchly test** and **compound symmetry**.

Hypergeometric distribution A **probability distribution** associated with **sampling without replacement** from a population of finite size N.

Mathematical details

If the population consists of r elements of one kind and $N - r$ of another, then the probability of finding x elements of the first kind when a **random sample** of size n is drawn is given by

$$\Pr(x) = \frac{\binom{r}{x}\binom{N-r}{n-x}}{\binom{N}{n}}$$

where Pr denotes probability, the mean of x is nr/N and its variance is

$$\frac{(nr)}{N}\left(1 - \frac{r}{n}\right)\left(\frac{N-n}{N-1}\right)$$

When N is large and n is small compared with N, the hypergeometric distribution can be approximated by the **binomial distribution**.

Numerical example

Suppose a population of 20 people consists of 15 psychologists and 5 statisticians. If 10 people are selected at random, what is the probability that there is a single statistician?

$$\Pr(\text{one statistician}) = \frac{\binom{5}{1}\binom{15}{9}}{\binom{20}{10}} = \frac{\frac{5!}{1!\,4!} \times \frac{15!}{9!\,6!}}{\frac{20!}{10!\,10!}} = 0.135$$

Hypothesis testing A general term for the procedure of assessing whether sample data are consistent or otherwise with statements made about the population. See also **null hypothesis**, **alternative hypothesis**, **composite hypothesis**, **significance test**, **significance level**, **Type I** and **Type II error**.

I

IDA Abbreviation for **initial data analysis**.

Identification The degree to which there is sufficient information in the sample observations to estimate the parameters in a proposed model. An *unidentified model* is one in which there are too many parameters in relation to the number of observations to make estimation possible. A *just identified model* corresponds to a **saturated model**. Finally an *overidentified model* is one in which parameters can be estimated, and there remain some degrees of freedom to allow the fit of the model to be assessed.

Example

Suppose the **correlation matrix** given below arises from a sample of examination scores for three subjects: Mathematics, French and English

$$R = \begin{bmatrix} 1.00 & & \\ r_{12} & 1.00 & \\ r_{13} & r_{23} & 1.00 \end{bmatrix}$$

A single-factor model is postulated for the correlations

$$x_1 = \lambda_1 f + u_1$$

$$x_2 = \lambda_2 f + u_2$$

$$x_3 = \lambda_3 f + u_3$$

Assuming that f is uncorrelated with the us then there are seven parameters to be estimated from the data; $\text{var}(f)$, $\text{var}(u_1)$, $\text{var}(u_2)$, $\text{var}(u_3)$, λ_1, λ_2 and λ_3. There are, however, only six sample statistics available for estimation: $\text{var}(x_1)$, $\text{var}(x_2)$, $\text{var}(x_3)$, correlation (x_1, x_2), correlation (x_1, x_3) and correlation (x_2, x_3). The model is underidentified and of no use as it stands.

Now assume that the latent variable f is in standardized form with $\text{var}(f) = 1$. Now there are six parameters to estimate with six sample statistics. The model is just identified and parameter estimates can now be found such that the predicted correlation matrix is identical to the observed correlation matrix, i.e. a perfect fit. But since the model contains exactly the same number of parameters as sample statistics, it does not represent a more parsimonious description of the data.

Lastly, we might assume that $\lambda_1 = \lambda_2 = \lambda_3$; the model now has only four parameters and is overidentified. If it provided an adequate fit for the observed correlations, it represents a parsimonious description. These are essentially the only models of practical importance.

Identity matrix A **diagonal matrix** in which all the elements on the leading diagonal (west to east) are unity and all other elements are zero. For example:

$$I - \begin{bmatrix} 1 & 0 & 0 \\ 0 & 1 & 0 \\ 0 & 0 & 1 \end{bmatrix}$$

Imputation A process for estimating **missing values** using the non-missing information available for a subject. Many methods have been developed, some of which involve the use of **multiple regression**. See also **last observation carried forward**.

Incidence A measure of the rate at which people without a particular condition develop the condition during a specific period of time. Calculated as incidence = number of new cases of a condition over a period of time ÷ population at risk of the condition in the time period. Measures the appearance of the condition. See also **prevalence**.

Incomplete contingency tables **Contingency tables** containing **structural zeros**. Such tables require special techniques for their analysis. See **Quasi-independence** .

Numerical example

Data showing health problems amongst a sample of teenagers

	Males		**Females**	
	12–15 yrs	**16–17 yrs**	**12–15 yrs**	**16–17 yrs**
Reproduction problems	4	2	9	7
Menstrual problems	–	–	4	8
How healthy I am	42	7	19	10
Nothing	57	20	71	31

Since males are not affected by menstrual problems certain cells are a priori zero.

Independence Essentially, two events are said to be independent if knowing the outcome of one tells us nothing about the other.

Mathematical details

Independence is defined formally in terms of the probabilities of the two events. In particular, two events A and B are said to be independent if

$$\Pr(A \text{ and } B) = \Pr(A) \times \Pr(B)$$

where $\Pr(A)$ and $\Pr(B)$ represent the probabilities of A and B.

See also **conditional probability** and **Bayes' Theorem**.

Independent samples *t*-test See **Student's *t*-test**.

Independent variables See **explanatory variables**.

Index plot A plot of some diagnostic quantity, such as a **residual**, obtained after the fitting of some model, against the corresponding observation number. Particularly suited to the detection of **outliers**.

Numerical example

The following data give the heights and weights of ten 11-year-old girls:

Child	Height (cm)	Weight (kg)
1	135	26
2	146	33
3	153	55
4	154	50
5	139	32
6	131	25
7	149	44
8	137	31
9	143	36
10	146	35

The estimated **linear regression** model for predicting weight from height is

$$\text{weight} = -133.91 + 1.191 \times \text{height}$$

The predicted weights of the ten girls and the difference between predicted and observed weights are:

Child	Predicted weight (kg)	Residual
1	26.81	−0.82
2	39.91	−6.91
3	48.25	6.75
4	49.44	0.56
5	31.58	0.42
6	22.06	2.94
7	43.49	0.51
8	29.20	1.80
9	36.34	−0.34
10	39.91	−4.91

The index plot of the residuals is shown in Figure 31. Children 2, 3 and 10 give some cause for concern.

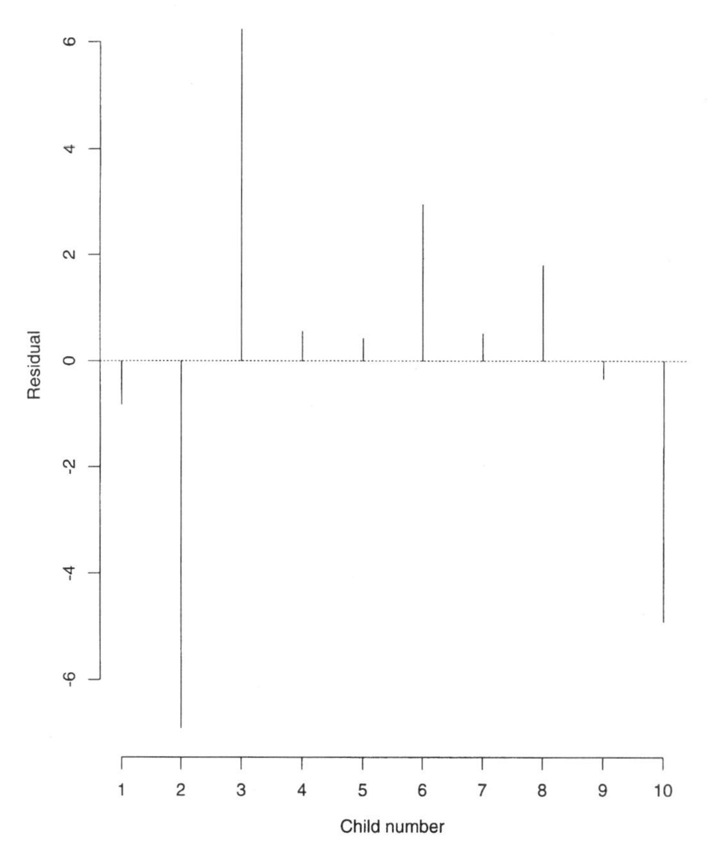

Figure 31 Index plot.

Indicator variable A term generally used for an observed variable (often also referred to as a **manifest variable**) when it is thought to be related to an underlying **latent variable** in the context of **structural equation models**.

Individual differences scaling (INDSCAL) A form of **multidimensional scaling** that allows for individual differences in the perception of the stimuli, allowing both an overall group stimulus space and private spaces or personal spaces for each individual. Used, for example, in situations where each of a number of individuals judges the similarities or dissimilarities of each pair of the same set of stimuli. [*Journal of Environmental Psychology*, 1996, 16, 75–92].

INDSCAL Acronym for **individual differences scaling**. [*Journal of Environmental Psychology*, 1996, 16, 75–92].

Inference The process of drawing conclusions about a population on the basis of measurements or observations made on a sample of individuals from the population. See also **frequentist inference** and **Bayesian inference**.

Influence statistics A range of statistics designed to assess the effect or influence of an observation in determining the results of a **regression analysis**. The general approach adopted is to examine the changes that occur in the **regression coefficients** when the observation is omitted. All such statistics can be computed from the results of the single regression using all data.

Influential observation An observation that has a disproportionate influence on one or more aspects of the estimate of a parameter, in particular, a **regression coefficient**. This influence may be due to differences from other subjects on the explanatory variables, an extreme value for the response variable, or a combination of these. **Outliers**, for example, are often also influential observations.

Informative missing values See **missing values**.

Information theory A branch of applied probability theory applicable to many communication and signal-processing problems in engineering and biology. Information theorists devote their efforts to quantitative examination of the following three questions:

- What is information?
- What are the fundamental limitations on the accuracy with which information can be transmitted?
- What design methodologies and computational algorithms yield practical systems for communication and storing information that perform close to the fundamental limits mentioned previously?

[*Ergonomics*, 1998, 41, 286–301].

Initial data analysis (IDA) The first phase in the examination of a data set, which consists of a number of informal steps, including:

- checking the quality of the data;
- calculating simple summary statistics and constructing appropriate graphs.

The general aim is to clarify the structure of the data, obtain a simple descriptive summary, and perhaps get ideas for a more sophisticated analysis.

Inliars A factitious term for **inliers**.

Inliers A term used for the observations most likely to be subject to error in statistics where a dichotomy is formed by making a 'cut' on an ordered scale, and where errors of classification can be expected to occur with greatest frequency in the neighbourhood of the cut. Suppose, for example, that individuals are classified on a 100-point scale that indicates degree of illness. A cutting point is chosen on the scale to dichotomize individuals into 'well' and 'ill' categories. Errors of classification are certainly more likely to occur in the neighbourhood of the cutting point.

Instantaneous death rate Synonym for **hazard rate**.

Instrumental variable A variable corresponding to an explanatory variable, x_1, that is correlated with x_1, but has no effect on the response variable except indirectly through x_1. Such variables are useful in deriving unbiased estimates of **regression coefficients** when the explanatory variables contain **measurement error**.

Integrated hazard function Synonym for **cumulative hazard function**.

Intention-to-treat analysis A procedure in which all participants randomly allocated to a treatment in a **clinical trial** are analysed together as representing that treatment, whether or not they completed or even received it. Here the initial random allocation not only decides the allocated treatment, but also decides there and then how the participant's data will be analysed, whether or not the participant actually receives the prescribed treatment. This method is adopted to prevent disturbances to the prognostic balance achieved by randomization and to prevent possible **bias** from using compliance, a factor often related to outcome, to determine the groups for comparison. [*Pain*, 1998, 74, 297–306].

Interaction A term applied when two (or more) explanatory variables do not act independently on a response variable. See also **additive effect**.

Intercept The parameter in an equation derived from a **regression analysis** corresponding to the **expected value** of the response variable when all the explanatory variables are zero.

Interpolation The process of determining a value of a function between two known values without using the equation of the function itself.

Inter-quartile range A measure of spread given by the difference between the first and third quartiles of a sample.

Interrupted time series design A study in which a single group of participants is measured several times before and after some event or manipulation. Often also used to describe investigations of a single individual. See also **longitudinal data** and ***N* of 1 clinical trial**.

Interval estimation See **estimation**.

Interval scale A scale established by measurement of a continuous variable. Differences on any part of the scale are equivalent. An example is Celsius temperature.

Interval variable Synonym for **continuous variable**.

Intervention study Synonym for **clinical trial**.

Interviewer bias The **bias** that occurs in surveys of human populations because of the direct result of the action of the interviewer. This bias can arise for a variety of reasons, including failure to contact the right persons and systematic errors in recording the answers received from the respondent.

Intra-class correlation Although originally introduced in genetics to judge sib-ship correlations, the term is now most often used for the proportion of variance of an observation due to between subject variability in the 'true' scores of a measuring instrument.

Mathematical details

If an observed value, x, is considered to be true score (t) plus measurement error (e), i.e.

$$x = t + e$$

the intra-class correlation is

$$\frac{\sigma_t^2}{(\sigma_t^2 + \sigma_e^2)}$$

where σ_t^2 is the variance of t and σ_e^2 the variance of e. The correlation can be estimated from a study involving a number of raters giving scores to a number of patients.

[*Pain*, 1997, 73, 253–257].

Intrinsically non-linear See **non-linear model**.

Invariance A property of a set of variables or a statistic that is left unchanged by a transformation. The variance of a set of observations is, for example, invariant under **linear transformations** of the data.

Item non-response A term used about data collected in a survey to indicate that particular questions in the survey attract refusals or responses that cannot be coded. Often this type of **missing data** makes reporting of the overall response rate for the survey less relevant. See also **non-response**.

Iteration The successive repetition of a mathematical process, using the result of one stage as the input for the next. An example of a procedure which involves iteration is **iterative proportional fitting**.

Iterative proportional fitting A procedure for the **maximum likelihood estimation** of the **expected frequencies** in **log-linear models**, particularly for models where such estimates cannot be found directly from simple calculations using relevant **marginal totals.**

J

Jaccard coefficient A similarity coefficient for use with data consisting of a series of binary variables in which 'negative' matches are not included in the calculation. See also **matching coefficient**.

Jackknife A procedure for reducing bias in estimation and providing approximate confidence intervals in situations where these are difficult to achieve in the usual way. The principle behind the method is to omit each sample member in turn from the data, thereby generating n separate samples each of size $n - 1$. The parameter of interest is then estimated from each sample and the estimates combined. Frequently used in discriminant analysis, for the estimation of the proportion of individuals misclassified by the derived classification rule. Calculated on the sample from which the rule is derived, the misclassification rate estimate is known to be optimistic. A jackknifed estimate obtained from calculating the discriminant function n times on the original observations, each time with one of the values removed, and in each case using it to classify the missing observations, usually gives a far more realistic estimate of the true misclassification rate. [*Journal of Experimental Education,* 1996, 64, 240–266].

Jittering A procedure for clarifying scatter diagrams when there is a multiplicity of points at many of the plotting locations, by adding a small amount of random variation to the data before graphing. An example involving the heights of both husbands and wives in 168 couples is shown in Figure 32.

Joint distribution Essentially synonymous with multivariate distribution, although used particularly as an alternative to bivariate distribution when two variables are involved.

Jonckheere's *k*-sample test A distribution-free method for testing the equality of a set of location parameters against an ordered alternative hypothesis.

J-shaped distribution An extremely asymmetric distribution with its maximum frequency in the initial (or final) class and a declining or increasing frequency elsewhere. An example is shown in Figure 33.

Just identified model See **identification**.

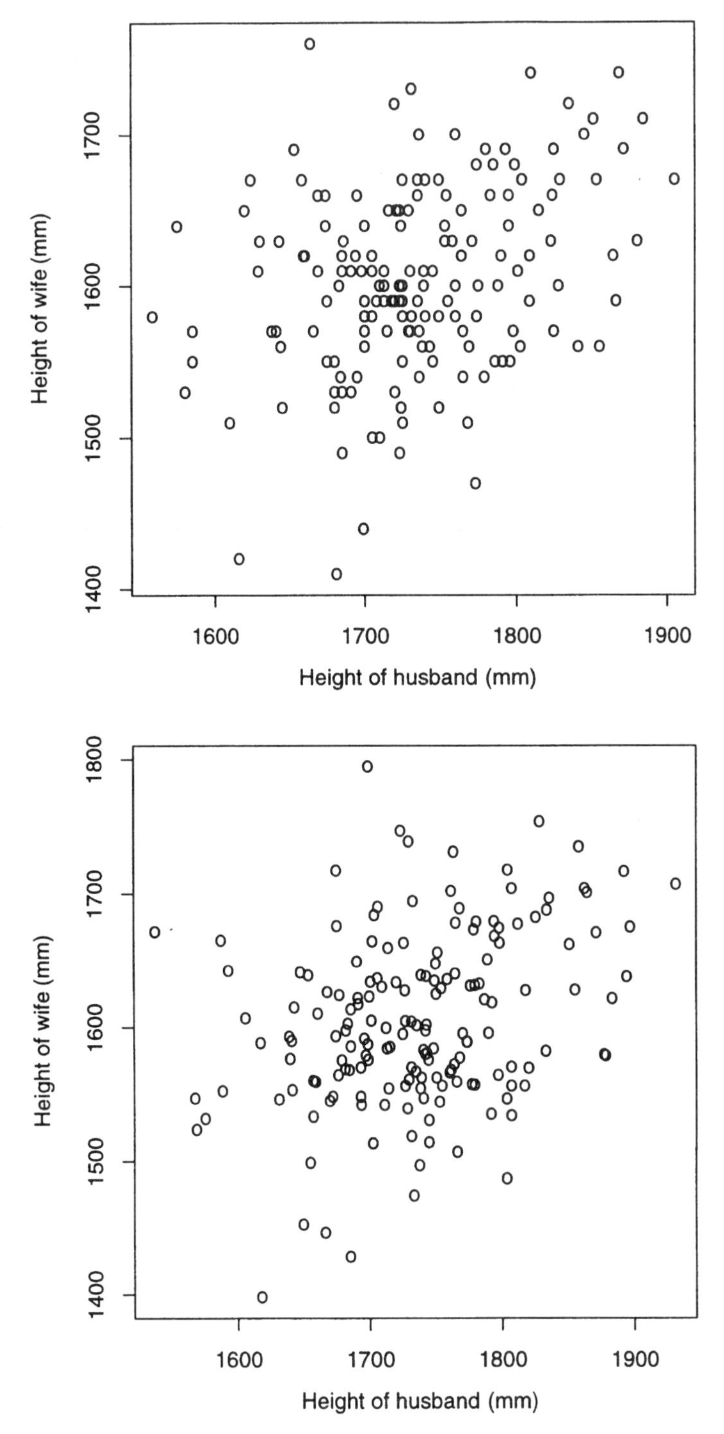

Figure 32 An illustration of jittering: top, raw data; bottom, data after jittering.

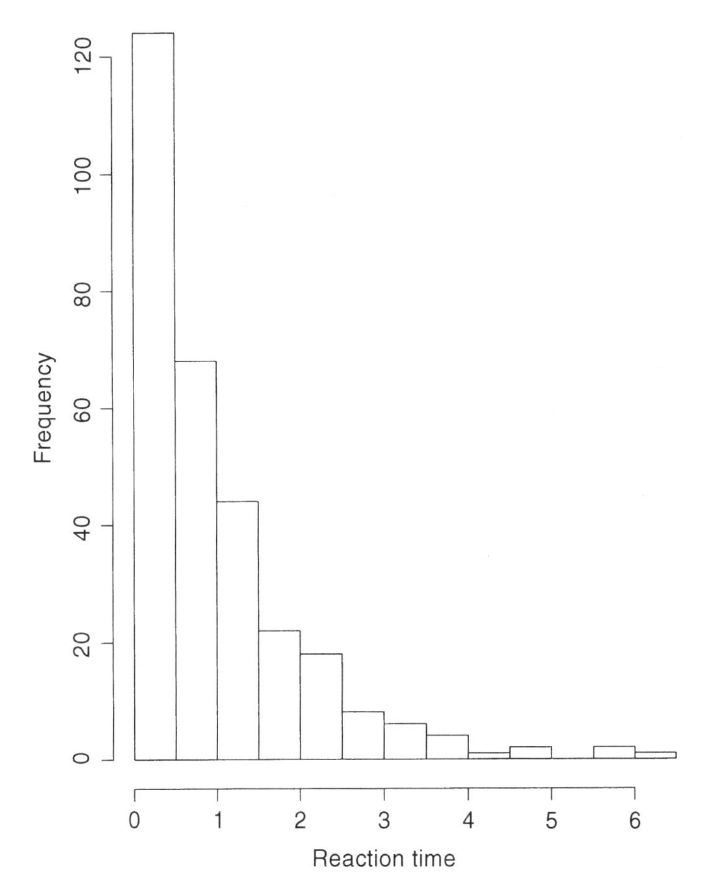

Figure 33 J-shaped distribution of reaction times.

K

Kaiser's rule A rule often used in **principal components analysis** for selecting the appropriate number of components. When the components are derived from the **correlation matrix** of the observed variables, the rule suggests retaining only those components with variances (**eigenvalues**) greater than one. See also **scree plot**.

Kaplan–Meier estimator Synonym for **product limit estimator**.

Kappa statistic A measure of the chance-corrected agreement between two observers in rating a categorical variable.

Mathematical details

Suppose the rating exercise on N subjects is summarized by the following table:

		Observer 1				
	Category	**1**	**2**	...	c	**Total**
	1	n_{11}	n_{12}		n_{1c}	$n_{1.}$
	2	n_{21}	n_{22}			$n_{2.}$
Observer 2	⋮					
	c	n_{c1}			n_{cc}	$n_{c.}$
	Total	$n_{.1}$			$n_{.c}$	N

The Kappa statistic is defined as

$$K = \frac{P_0 - P_c}{1 - P_c}$$

where P_0 is the observed proportion of agreement given by

$$P_0 = \frac{n_{11} + n_{22} + \cdots + n_{cc}}{N}$$

and P_c is the agreement to be expected by chance if the observers rate according to their **marginal totals** in each category, given by

$$P_c = \frac{1}{N}\left[\frac{n_{1.} \times n_{.1}}{N} + \frac{n_{2.} \times n_{.2}}{N} + \cdots + \frac{n_{c.} \times n_{.c}}{N}\right]$$

The Kappa statistic gives the ratio of the excess of observed agreement over chance agreement to the maximum possible excess.

Numerical example

The data below show the ratings of two observers for 100 subjects on a variable with three categories:

		Observer 1			
		1	**2**	**3**	**Total**
	1	24	13	3	40
Observer 2	**2**	5	20	5	30
	3	1	7	22	30
	Total	30	40	30	100

$$P_0 = 66/100 = 0.66$$

$$P_c = \frac{1}{100}\left[\frac{40 \times 30}{100} + \frac{30 \times 40}{100} + \frac{30 \times 30}{100}\right] = 0.33$$

$$K = \frac{0.66 - 0.33}{1 - 0.33} = 0.49$$

Such a value indicates only moderate agreement between the two observers.

See also **weighted kappa**.

Kendall's coefficient of concordance Synonym for **coefficient of concordance**.

Kendall's tau A **rank correlation coefficient** for two variables.

Mathematical details

The coefficient is based on the number of inversions in one ranking as compared with the other, that is on S, given by

$$S = P - Q$$

where P is the number of concordant pairs of observations – that is, pairs of observations such that their rankings on the two variables are in the same direction – and Q is the number of discordant pairs for which rankings on the two variables are in the reverse direction.

The coefficient is given by

$$\tau = \frac{2S}{n(n-1)}$$

where n is the number of observations.

Numerical example

The following data give the ranks of 10 students in French and German exams, with the ranks for French being put into order.

	Student									
	A	I	H	E	D	G	F	C	J	B
French rank	1	2	3	4	5	6	7	8	9	10
German rank	1	3	2	7	5	4	8	6	9	10

So, we have $P = 9 + 7 + 7 + 3 + 4 + 4 + 2 + 2 + 1 = 39$ and $Q = 6$, $S = 33$ and $\tau = 0.733$. This indicates that there is a relatively strong relationship between the ratings on the two subjects.

See also **phi-coefficient**, **Cramér's V** and **contingency coefficient**.

K-means cluster analysis A method of cluster analysis which seeks a partition of the n individuals into a specified number of groups so as to optimize some numerical criterion which characterizes the quality of the solution. For univariate data, for example, selecting a partition to maximize the between groups sum of squares or, equivalently, minimizing the within groups sum of squares might be used. Analogous criteria are available for multivariate data. See also **agglomerative hierarchical clustering methods**. [*International Journal of Organizational Analysis*, 1996, 4, 20–51].

Kolmogorov–Smirnov two-sample method A distribution-free method that tests for any difference between two population probability distributions. The test is based on the maximum absolute difference between the cumulative distribution functions of the samples from each population. Critical values are available in many statistical tables.

Kruskal–Wallis test A distribution-free method that is the analogue of the analysis of variance for a one-way design. It tests whether the k groups to be compared have the same population median.

Mathematical details

The test statistic is derived by ranking all the N observations from 1 to N regardless of which group they are in, and then calculating

$$H = \frac{12 \sum_{i=1}^{k} n_i (\bar{R}_i - \bar{R})^2}{N(N-1)}$$

where n_i is the number of observations in group i, $\bar{R}_i$ is the mean of their ranks, $\bar{R}$ is the average of all the ranks, given explicitly by $(N + 1)/2$. When the null hypothesis is true, the test statistic has a chi-squared distribution with $k - 1$ degrees of freedom.

Numerical example

An experimenter wishes to test whether or not three type styles have identical effects on reading speed. Five subjects are randomly allocated to each type style. Each subject's reading speed is measured. The results are as follows:

Type style		
1	**2**	**3**
135 (8)	175 (11)	105 (3)
91 (2)	130 (7)	147 (9)
111 (5)	514 (15)	159 (10)
87 (1)	283 (13)	107 (4)
122 (6)	301 (14)	194 (12)

(Ranks are shown in parentheses.)

$$H = \frac{12 \times 5[(4.4 - 8)^2 + (12 - 8)^2 + (7.2 - 8)^2]}{15(15 - 1)} = 8.46$$

Testing as a chi-squared with 2 df, H has an associated **P-value** less than 0.05, indicating that there is a difference in reading speed for the three type styles.

[*Journal of Sport and Exercise Psychology*, 1997, 19, 156–168].

Kurtosis The extent to which the peak of a unimodal **probability distribution** or **frequency distribution** departs from the shape of the peak of a **normal distribution**; if the peak is more pointed than a normal curve, the term *leptokurtic* is used; if flatter, the term *platokurtic* is used. Examples of each type are shown in Figure 34.

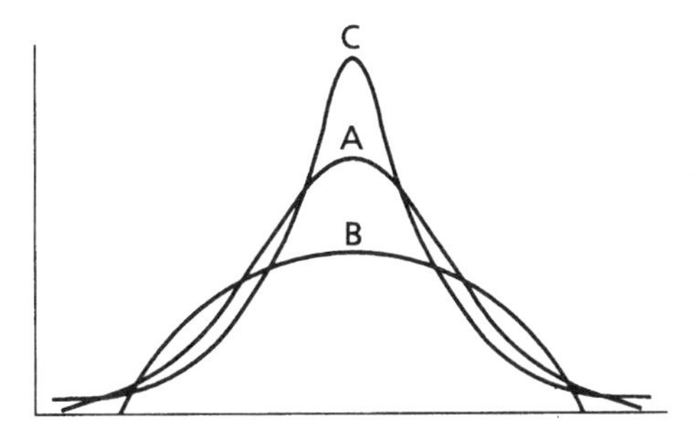

Figure 34 Illustrations of kurtosis: A = normal, B = platokurtic, C = leptokurtic.

L

Large-sample method Any statistical method based on an approximation to a **normal distribution** or other **probability distribution** that becomes more accurate as sample size increases.

Last observation carried forward (LOCF) A method for replacing the observations of participants who drop out of a **clinical trial** carried out over a period of time. It consists of substituting for each **missing value** the participant's last available assessment of the same type. Although widely applied, particularly in the pharmaceutical industry, its usefulness is very limited since it makes very unlikely assumptions about the data: for example, that the values are frozen at their last recorded value. See also **imputation**.

Latent class analysis A method of analysis useful for exploring the structure of categorical data in particular data sets consisting only of **binary variables**. The model fitted assumes that the associations between the observed variables are a result of the existence of different groups of observations within which the variables are independent. Parameters in such models can be estimated by **maximum likelihood estimation**. Can be considered as either an analogue of **factor analysis** for categorical variables or a model of **cluster analysis** for such data. [*Psychological Medicine*, 1997, 27, 835–845].

Latent variable A variable that cannot be measured directly, but is assumed to be related to a number of observable or **manifest variables**. Possible examples include racial prejudice and social class. See also **indicator variable**.

Latin square An **experimental design** aimed at removing the variation from two extraneous sources from the experimental error, so that a more sensitive test of the treatment effect can be achieved. The rows and columns of the square represent the levels of the two extraneous factors. The treatments are represented by roman letters so that no letter appears more than once in each row and column. It is assumed that there are no interactions between the three factors. The design achieves a considerable saving in number of subjects when compared with the corresponding complete factorial design. As an example consider an experiment in which four teaching methods, A, B, C and D, are given to students of four levels of ability from four different schools. A possible 4×4 Latin square design would be:

	L1	L2	L3	L4
S1	A	B	C	D
S2	B	C	D	A
S3	C	D	A	B
S4	D	A	B	C

So students from school 1 in level 1 are given A etc. Upon completion of the experiment, treatment A will have been administered to some students from each of the ability levels as well as some students from each of the schools. This same kind of balance holds for the other treatments. [*Journal of the Experimental Analysis of Behaviour*, 1998, 69, 77–85].

Leaps-and-bounds algorithm An **algorithm** used to find the optimal solution in problems that have a very large number of possible solutions. Begins by splitting the possible solutions into a number of exclusive subsets, and limits the number of subsets that need to be examined in searching for the optimal solution by a number of different strategies. Often used in **all subsets regression** to restrict the number of models that have to be examined.

Least significant difference (LSD) test An approach to comparing a set of means that controls the **familywise error rate** at some particular level, say α. The hyothesis of the equality of means is tested first by an α-level **F-test**. If this test is not significant, then the procedure terminates without making detailed inferences on pairwise differences; otherwise each pairwise difference is tested by an α-level **Student's t-test**.

Least squares estimation A method used for estimating parameters of some model of interest, which operates by minimizing the squared difference between the observed response and the value predicted by the model.

Mathematical details

Consider the simple linear regression model

$$y_i = \beta_0 + \beta_1 x_i + \varepsilon_i$$

The least squares estimators of the parameters β_0 and β_1 are obtained from n pairs of sample values, $(x_1, y_1), (x_2, y_2), \ldots, (x_n, y_n)$ by minimizing S, given by

$$S = \sum_{i=1}^{n} (y_i - \beta_0 - \beta_1 x_i)^2$$

leading to

$$\hat{\beta}_0 = \bar{y} - \hat{\beta}\,\bar{x}$$

$$\hat{\beta} = \frac{\sum_{i=1}^{n} (x_i - \bar{x})(y_i - \bar{y})}{\sum_{i=1}^{n} (x_i - \bar{x})^2}$$

For a numerical example, see **index plot**.

Often referred to as *ordinary least squares,* to differentiate this simple version of the technique from more involved versions, such as **weighted least squares**.

Leptokurtic See **kurtosis**.

Leverage points A term used in **regression analysis** for those observations that have an extreme value on one or more explanatory variables. The effect of

; test

is to force the fitted model close to the observed value of the response, small **residual**. See also **hat matrix**.

A test for equality of variances that is relatively insensitive to ... from normality. See also **Bartlett's test** and **Hartley's test**.

Lie factor A quantity sometimes used for judging the honesty of a graphical presentation of data. Calculated as

$$\frac{\text{apparent size of effect shown in graph}}{\text{actual size of effect in data}}$$

Values close to unity are desirable, but it is not uncommon to find values close to zero and greater than five. The example shown in Figure 35 has a lie factor of about 2.8.

Figure 35 A diagram with a lie factor of 2.8.

Life expectancy The expected number of years remaining to be lived by persons of a particular age. For example, according to the US life table for 1979–1981, the life expectancy at birth is 73.88 years and that at age 40 is 36.79 years. The life expectancy of a population is a general indication of the capability of prolonging life. It is used to identify trends and to compare longevity.

Likelihood The probability of a set of observations given the value of some parameter or set of parameters. The function is the basis of **maximum likelihood estimation**.

Likelihood ratio The ratio of the **likelihoods** L_{H_0} and L_{H_1} of the data under two hypotheses, H_0 and H_1. Can be used to assess H_0 against H_1 since, under H_0, the statistic, λ, given by

$$\lambda = -2 \ln \frac{L_{H_0}}{L_{H_1}}$$

has approximately a **chi-squared distribution** with degrees of freedom equal to the difference in the number of parameters in the two hypotheses. See also **deviance** and **goodness-of-fit statistics**.

Likert scales Scales often used in studies of attitudes, in which the raw scores are based on graded alternative responses to each of a series of questions. For example, the subject may be asked to indicate his/her degree of agreement with each of a series of statements relevant to the attitude. A number is attached to each possible response – e.g. 1, strongly approve; 2, approve; 3, undecided; 4, disapprove; 5, strongly disapprove – and the sum of these used as the composite score. A commonly used example in psychology would be anxiety levels of an agoraphobic patient in different situations perhaps measured by the Spielberger State Anxiety Inventory.

Linear-by-linear association test A test for detecting specific types of departure from independence in a **contingency table** in which both the row and column categories have a natural order. See also **chi-squared test for trend**.

Linear estimator An estimator which is a **linear function** of the observations, or of sample statistics calculated from the observations.

Linear function A function of a set of variables, parameters, etc., that does not contain powers or cross-products of the quantities.

Mathematical details

For example, the following are all such functions of three variables, x_1, x_2 and x_3:

$$y = x_1 + 2x_2 + x_3$$

$$z = 6x_1 - x_3$$

$$w = 0.34x_1 - 2.4x_2 + 12x_3$$

Linear logistic regression Synonym for **logistic regression**.

Linear model A model in which the **expected value** of a **random variable** is expressed as a **linear function** of the parameters in the model.

Mathematical details

Examples of linear models are

$$E(y) = \alpha + \beta x$$

$$E(y) = \alpha + \beta x + \gamma x^2$$

where x and y represent variable values and α, β and γ parameters. Note that the linearity applies to the parameters, not to the variables.

See also **linear regression** and **generalized linear models**.

Linear regression A term usually reserved for the simple **linear model** involving a response, y, that is a continuous variable and a single explanatory variable, x, related by the equation

$$E(y) = \alpha + \beta x$$

where E denotes the **expected value**. See also **multiple regression** and **least squares estimation.**

Linear transformation A transformation of q variables, $x_1, x_2, \ldots, x_q$, given by the q equations

$$y_1 = a_{11} x_1 + a_{12} x_2 + \cdots + a_{1q} x_q$$

$$y_2 = a_{21} x_1 + a_{22} x_2 + \cdots + a_{2q} x_q$$

$$\vdots$$

$$y_q = a_{q1} x_1 + a_{q2} x_2 + \cdots + a_{qq} x_q$$

Such a transformation is the basis of **principal components** analysis.

Linear trend A relationship between two variables in which the value of one changes at a constant rate as the other increases.

Link function See **generalized linear models.**

LISREL A computer program for fitting **structural equation models** involving **latent variables**. See also **EQS**.

Loading matrix See **factor analysis.**

Location The notion of central or 'typical value' in a sample distribution. See also **mean**, **median** and **mode**.

LOCF Abbreviation for **last observation carried forward.**

Logarithmic transformation The transformation of a variable, x, obtained by taking $y = \ln(x)$. Often used when the **frequency distribution** of the variable, x, shows a moderate to large degree of **skewness**, in order to achieve normality.

Logistic regression A form of **regression analysis** used when the dependent variable takes values of only zero and one.

Mathematical details

One possible regression model for **binary variables** would involve a **linear model** for the probability that the dependent variable takes the value one, i.e.

$$p = \beta_0 + \beta_1 x_1 + \beta_2 x_2 + \cdots + \beta_q x_q$$

where $x_1, \ldots, x_q$ are the explanatory variables of interest. Such a model could be fitted using **least squares estimation**, but there are two serious drawbacks:

(a) the approach could lead to fitted values of p outside the range $(0, 1)$;
(b) the assumption of a **normal distribution** for the dependent variable is clearly not justified for binary variables.

Consequently the model used in logistic regression is

$$\log \frac{p}{1-p} = \beta_0 + \beta_1 x_1 + \cdots + \beta_q x_q$$

When p tends to 0, $\log p/(1-p)$ tends to $-\infty$, and when p tends to 1, $\log p/(1-p)$ tends to ∞, thus removing the previous constraint in the response that is modelled. Parameters in the model are estimated by **maximum likelihood estimation**.

Numerical example

Suppose we are interested in assessing the relationship between whether or not a person is depressed and their sex and age from the following small set of data:

Depressed (1 = yes, 0 = no)	Sex (1 = M, 0 = F)	Age (years)
0	1	27
0	1	35
1	1	40
1	0	28
1	0	49
1	0	50
0	0	24
0	1	21
0	1	23
1	0	52
0	1	27
0	1	28
0	1	33
0	1	41
0	0	20

The estimated parameters in a logistic regression and their standard errors are:

	Parameter estimate	SE
Sex	13.34	12.3
Age	−0.92	0.83
Constant	23.95	21.7

Comparing each estimated regression coefficient with its standard error indicates that neither sex nor age is significantly related to being depressed.

[*Journal of Experimental Child Psychology*, 1998, 68, 51–69].

Log-likelihood The logarithm of the likelihood. Generally easier to work with than the likelihood itself when using maximum likelihood estimation.

Log-linear models Models for count data in which the logarithm of the expected value of a count variable is modelled as a linear function of parameters; the latter represent associations between pairs of variables and higher order interactions between more than two variables. Estimated expected frequencies under particular models are found from iterative proportional fitting. Such models are, essentially, the equivalent, for frequency data, of the models for continuous data used in analysis of variance, except that interest usually now centres on parameters representing interactions, rather than those for main effects. See also **generalized linear model**, and **Poisson regression**. [*Journal of Occupational and Organizational Psychology*, 1994, 67, 259–278].

Logrank test A method for comparing the survival times of two or more groups of subjects that involves the calculation of observed and expected frequencies of failures in separate time intervals. The relevant test statistic is, essentially, a comparison of the observed number of deaths occurring at each particular time point with the number to be expected if the survival experience of the two groups is the same.

Numerical example

Survival times for males and females

Patient	Sex	Survival time (weeks)	Status at end of study
1	M	2.3	Dead
2	M	4.8	Alive
3	M	6.1	Dead
4	M	15.2	Dead
5	M	23.8	Alive
6	F	1.6	Dead
7	F	3.8	Dead
8	F	14.3	Alive
9	F	18.7	Dead
10	F	36.3	Alive

Calculation of logrank test for data

Time	Status	M	F	Total
1.6	Dead	0(0.5)	1(0.5)	1
	Alive	5	4	9
	Total	5	5	10
2.3	Dead	1(0.55)	0(0.45)	1
	Alive	4	4	8
	Total	5	4	9
3.8	Dead	0(0.5)	1(0.5)	1
	Alive	4	3	7
	Total	4	4	8
6.1	Dead	1(0.5)	0(0.5)	1
	Alive	2	3	5
	Total	3	3	6
15.2	Dead	1(0.5)	0(0.5)	1
	Alive	1	2	3
	Total	2	2	4
18.7	Dead	0(0.33)	1(0.67)	1
	Alive	1	1	2
	Total	1	2	3

Expected number of deaths are shown in parentheses. 'Alive' means alive *and* at risk. Observed number of deaths for men = 3. Expected number of deaths for men = 0.5 + 0.56 + 0.5 + 0.5 + 0.5 + 0.33 = 2.89.

Observed number of deaths for women = 3. Expected number of deaths for women = 0.5 + 0.44 + 0.5 + 0.5 + 0.5 + 0.67 = 3.11.

$$\chi^2 = \frac{(3 - 2.89)^2}{2.89} + \frac{(3 - 3.11)^2}{3.11} = 0.004 + 0.004 = 0.008$$

Indicates that there is no difference in the survival experience in the two groups.

LOGXACT A specialized statistical package that provides exact inference capabilities for **logistic regression**.

Longitudinal data Data arising from studies in which a dependent variable is recorded on each participant at a number of different times. Such data combine elements of **multivariate data** and **time series** data. They differ from the former, however, in that only a single variable is involved, and from the latter in consisting of a (possibly) large number of short series, one from each participant, rather than a single long series. Such data can be collected either prospectively, following participants forward in time, or retrospectively, by extracting measurements on each person from historical records. This type of data is also often known as *repeated-measures data,* particularly in the social and behavioural sciences, although in these disciplines such data are more likely to arise from observing individuals repeatedly under different experimental conditions, rather than from a simple time sequence. Special statistical methods are often needed for the analysis of this type of data because the set of measurements on one participant tend to be intercorrelated. This correlation must be taken into account to draw valid scientific inferences. See also **Greenhouse–Geisser correction**, **Huynh–Feldt correction**, **compound symmetry** and **Mauchly test**.

Longitudinal study A study that gives rise to longitudinal data. The defining characteristic of such a study is that participants are measured repeatedly through time.

Lower triangular matrix A matrix in which all the elements above the main diagonal are zero. Occurs in accounts of many methods of multivariate analysis. An example is the following:

$$\mathbf{L} = \begin{bmatrix} 1 & 0 & 0 & 0 \\ 2 & 3 & 0 & 0 \\ 1 & 1 & 3 & 0 \\ 1 & 5 & 6 & 7 \end{bmatrix}$$

LSD Abbreviation for **least significant difference**.

M

Main effect An estimate of the independent effect of (usually) a factor variable on a response variable in **analysis of variance**.

Mainframes High speed general purpose computers with a very large storage capacity.

Mallow's C_k statistic An index used in **regression analysis** as an aid in choosing the 'best' subset of **explanatory variables**. The model chosen is the one with the minimum value of the index.

Mathematical details

The index, C_k, is defined as

$$C_k = \sum_{i=1}^{n} \frac{(y_i - \hat{y}_i^{(k)})^2}{s^2} - n + 2q$$

where n is the number of observations, y_i is the observed value of the dependent variable for individual i, $\hat{y}_i^{(k)}$ is the corresponding predicted value from a fitted model with a particular set of k explanatory variables, and s^2 is the **residual mean square** after regression on the complete set of q explanatory variables.

Numerical example

The data below give the current annual income, the number of years of formal education, the number of years in the job and IQ of 10 employees of a large company:

Subject	Income	Years of education	Years in job	IQ
1	8 000	5	3	101
2	10 000	6	2	110
3	10 500	4	4	99
4	15 000	7	3	105
5	12 000	4	8	101
6	25 000	8	6	120
7	17 000	9	1	125
8	30 000	7	10	115
9	6 000	4	2	95
10	20 000	6	8	110

The following values of the C_k index can be calculated:

For subsets of one variable the values of the C_k statistic are

Variable	C_k
IQ	47.93
Years in job	55.53
Years of education	57.28

The best single predictor is IQ.

For subsets of two variables the values of the C_k statistic are

Variable	C_k
Years of education, Years in job	2.06
Years in job, IQ	8.85
Years of education, IQ	49.76

The best pair of predictors is years of education and years in job.

See also **selection methods in regression** and **all subsets regression**.

Manifest variable A variable that can be measured directly, in contrast to a latent variable.

Mann–Whitney test A distribution-free method used as an alternative to the Student's *t*-test for assessing whether two populations have the same location. Given a sample of observations from each population, all the observations are ranked as if they were from a single sample, and the test statistic is the sum of the ranks in the smaller group. Tables giving critical values of the test statistic are available, and a normal approximation can be used for moderate and large sample sizes.

Numerical example

A pharmaceutical firm markets a product that they claim will reduce the absorption of alcohol into the blood stream. To test the claim, nine volunteers are each asked to drink three pints of beer; four of the nine selected at random are given the product to take. After 30 minutes a blood sample is taken from each of the nine and the blood alcohol content in mg/100 ml for each is as follows:

- Given product: 79, 85, 105, 93
- Not given product: 99, 102, 107, 117, 108

The question of interest is whether the preparation influences the alcohol level in the blood stream. Firstly we jointly rank all the observations:

Observation	79*	85*	93*	99	102	105*	107	108	117
Rank	1	2	3	4	5	6	7	8	9

The sum of the ranks in the smaller sample (those given the product) is $1 + 2 + 3 + 6 = 12$. This leads to a **two tail test *P*-value** of approximately 0.06. Consequently there is insufficient evidence to conclude that the product influences alcohol level.

MANOVA Acronym for **multivariate analysis of variance**.

Marginal homogeneity A term applied to **square contingency tables** when the probabilities of falling in each category of the row variable equal the corresponding probabilities for the column variable.

Marginal matching The **matching** of treatment groups in terms of means or other summary characteristics of the matching variables. Has been shown to be almost as efficient as the matching of individual subjects in some circumstances.

Marginal totals A term often used for the total number of observations in each row and each column of a **contingency table**.

Markov chain A discrete **stochastic process** in which the probability that a system will be in a given state on the $(k + 1)$th trial depends only on the state of the system on the kth trial. [*Journal of Instructional Psychology*, 1996, 23, 245–248].

Matched case–control study See **retrospective study**.

Matched pairs *t*-test A **Student's *t*-test** for testing the equality of the means of two populations, when the observations arise as **paired samples**. The test is based on the differences between the observations of the matched pairs.

Mathematical details

The **test statistic**, t, is given by

$$t = \frac{\bar{d}}{s_d/\sqrt{n}}$$

where n is the sample size, $\bar{d}$ is the mean of the differences, and s_d is their standard deviation. If the null hypothesis of the equality of the population means is true, then t has a **Student's *t*-distribution** with $n - 1$ degrees of freedom.

Numerical example

The following data give the head length measurements (mm) for each of the first two adult sons in 10 families:

Head length of first son	Head length of second son
191	179
195	201
181	185
183	188
176	171
208	192
189	190
197	189
188	197
192	187

A matched t-test can be used to assess whether there is any evidence of a difference in head lengths of first and second sons:

$$\bar{d} = 2.10, \qquad \bar{s}_d = 8.36, \qquad n = 10$$

$$t = \frac{2.10 \times \sqrt{10}}{8.36} = 0.79$$

This is clearly non-significant so there is no evidence of a difference in the head lengths of first and second sons.

Matching The process of making a study group and a comparison group comparable with respect to extraneous factors. Often used in a retrospective study, when selecting cases and controls, to control variation in a response variable due to sources other than those immediately under investigation. Several kinds of matching can be identified, the most common of which is when a case is individually matched with a control subject on the matching variables, such as sex, occupation. When the variable on which the matching takes place is continuous (e.g. age), it is usually transformed into a series of categories but a second method is to say that two values of the variable match if their difference lies between defined limits. This method is known as *caliper matching*. See also **paired samples**.

Matching coefficient A similarity coefficient for data consisting of a number of binary variables that is often used in cluster analysis. Given by the number of variables on which two individuals match divided by the total number of variables.

Mauchly test A test that a variance–covariance matrix of a set of multivariate data is a scalar multiple of the identity matrix, a property known as *sphericity*. Of most importance in the analysis of longitudinal data, where this property must hold for the F-tests in the analysis of variance of such data to be valid. See also **compound symmetry**, **Greenhouse–Geisser correction** and **Huynh–Feldt correction**.

Maximum F-ratio Equivalent to Hartley's test. See also **Bartlett's test** and **Box's test**.

Maximum likelihood estimation (MLE) A procedure for finding estimates of parameters in many types of models. Such estimates can be shown to have many desirable statistical properties.

Mathematical details

The estimation procedure is based on the **likelihood** of the data, i.e. the probability of the data given the parameters. So, for example, if we have observations $x_1, x_2, \ldots, x_n$, which arise from a **probability distribution** $f(x; \theta)$ the likelihood is

$$L = \prod_{i=1}^{n} f(x_i; \theta)$$

The maximum likelihood estimator of θ would be found by maximizing L with respect to θ. (It is often easier to maximize the **log-likelihood** than the likelihood itself.)

As a specific example consider observations from an exponential distribution $f(x; \theta) = \theta\,\mathrm{e}^{-\theta x}$. The likelihood of a **random sample** of n observations is given by

$$L = \prod_{i=1}^{n} f(x_i; \theta) = \prod_{i=1}^{n} \theta\,\mathrm{e}^{-\theta x_i} = \theta^n \exp\left(-\theta \sum_{i=1}^{n} x_i\right)$$

The log-likelihood, $L = \log L$, is therefore

$$L = n \log \theta - \theta \sum_{i=1}^{n} x_i$$

To maximize L with respect to θ, we differentiate it with respect to θ and set the result to zero:

$$\frac{\mathrm{d}L}{\mathrm{d}\theta} = \frac{n}{\theta} - \sum x_i = 0$$

Consequently the maximum likelihood estimator of θ, $\hat{\theta}$, is given by

$$\hat{\theta} = \frac{n}{\sum_{i=1}^{n} x_i} = \frac{1}{\bar{x}}$$

the reciprocal of the sample mean.

Numerical example

The following are 10 observations generated from an exponential distribution with $\theta = 1$: 0.76, 1.32, 0.17, 0.24, 2.92, 1.80, 2.18, 0.02, 0.51, 0.05. The sample mean is 0.997, so the maximum likelihood estimator of θ is $1/0.997 = 1.003$.

[*Applied Psychological Measurement*, 1997, 21, 321–336].

MCAR Abbreviation for **missing completely at random**.

McNemar's test A test for comparing proportions in data involving **paired samples**.

Mathematical details

The **test statistic** is given by

$$\chi^2 = \frac{(b-c)^2}{b+c}$$

where b is the number of pairs for which the individual receiving treatment A has a positive response and the individual receiving treatment B does not, and c is the number of pairs for which the reverse is the case. If the probability of a positive response is the same in each group, then χ^2 has a **chi-squared distribution** with a single degree of freedom.

Numerical example

A psychologist wished to assess the effect of the symptom depersonalization on the progress of depressed patients. For this purpose 23 depressed patients who were diagnosed as being depersonalized were matched one-to-one on age, sex, duration of illness and certain personality variables, with 23 depressed patients who were diagnosed as not being depersonalized. After all patients had completed a course of cognitive behaviour therapy, they were assessed as recovered or not recovered with the following results:

	Depressed patients		
Patients not depersonalized	**Recovered**	**Not recovered**	**Total**
Recovered	14	5	19
Not recovered	2	2	4
Total	7	16	23

The McNemar test statistic is

$$\chi^2 = \frac{(5-2)^2}{5+2} = 1.3$$

This is not significant, and so depersonalization appears not to be associated with prognosis in patients with endogenous depression.

Mean A measure of location or central value for a continuous variable. For a definition of the population value see **expected value**. For a sample of observations, $x_1, x_2, \ldots, x_n$, the measure is calculated as

$$\bar{x} = \frac{\sum_{i=1}^{n} x_i}{n}$$

Most useful when the data have a **symmetric distribution** and do not contain **outliers**. Otherwise the median is preferable. See also **mode**, **geometric mean** and **harmonic mean**.

Mean square contingency coefficient The square of the **phi-coefficient**.

Mean square ratio The ratio of two **mean squares** in an **analysis of variance**.

Mean squares (MSE, MSe, MS_e) The name used in the context of **analysis of variance** for estimators of a particular variance of interest. For example, in the analysis of a **one-way design**, the **within groups mean square** estimates the assumed common variance in the k groups. This is often also referred to as the **error mean square**.

Mean vector A vector containing the mean values of each variable in a set of **multivariate data**.

> *Numerical example*
>
> Under the **Mallow's C_k statistic** entry a set of data for 10 individuals on four variables is given. The mean vector of this data is simply a vector of the means of each variable, i.e.
>
> $$[15\,350, 6.00, 4.70, 108.10]$$

Measurement error Errors in reading, calculating or recording a numerical value. The difference between the observed values of a variable recorded under similar conditions and some fixed true value.

Measures of association Numerical indices quantifying the strength of the statistical dependence of two or more qualitative variables. See also **phi-coefficient**.

Median The value in a set of ranked observations that divide the data into two parts of equal size. When there is an odd number of observations the median is the middle value. When there is an even number of observations the median is calculated as the average of the two central values. Provides a measure of location of a sample that is suitable for **asymmetric distributions** and is also relatively insensitive to the presence of **outliers**.

> *Numerical example*
>
> Consider the following set of 10 observations on IQ: 101, 108, 92, 115, 122, 97, 91, 110, 105, 112. The median is given by $\frac{1}{2}(105 + 108) = 106.5$.

See also **mean** and **mode**.

MEDLINE Medical Literature Analysis Retrieval System on line published by the US National Library of Medicine. The CDs contain medical literature from 1966. 3700 journals are covered from 70 countries.

Mesokurtic See **kurtosis**.

Meta-analysis A collection of techniques whereby the results of two or more independent studies are statistically combined to yield an overall answer to a question of interest. The rationale behind this approach is to provide a test with more **power** than is provided by the separate studies themselves. The procedure has become increasingly popular in the last decade or so, but it is not without its critics, particularly because of the difficulties of knowing which studies should be included and to which population the final results actually apply. See also **funnel plot** and **publication bias**. [*Brain and Cognition*, 1998, 36, 209–236].

Mid *P*-value An alternative to the conventional ***P*-value** that is used, in particular, in some analyses of discrete data. For example, **Fisher's exact test** on **two-by-two contingency tables**. In the latter, if $x = a$ is the observed value of the frequency of interest, and this is larger than the value expected, then the mid *P*-value is defined as

$$\text{mid } P\text{-value} = \tfrac{1}{2}\Pr(x = a) + \Pr(x > a)$$

In this situation, the usual *P*-value would be defined as $\Pr(x \geq a)$, where Pr denotes probability.

Mid-range The mean of the smallest and largest values in a sample. Sometimes used as a rough estimate of the mean of a **symmetric distribution**.

MIMIC model Abbreviation for **multiple indicator multiple cause model**.

MINITAB A general purpose statistical software package, specifically designed to be useful for teaching purposes.

Misinterpretation of *P*-values A ***P*-value** is commonly interpreted in a variety of ways that are incorrect. Most common are that it is the probability of the null hypothesis, and that it is the probability of the data having arisen by chance. For the correct interpretation, see the entry for ***P*-value**.

Missing at random (MAR) See **missing values**.

Missing completely at random (MCAR) See **missing values**.

Missing values Observations missing from a set of data for some reason. In **longitudinal studies**, for example, they may occur because participants drop out of the study completely or do not appear for one or other of the scheduled visits or because of equipment failure. Common causes of participants prematurely ceasing to participate include recovery, lack of improvement, unwanted effects that may be related to the investigation, unpleasant study procedures and intercurrent health problems. Such values greatly complicate many methods of analysis, and simply using those individuals for whom the data are complete can be unsatisfactory in many situations. A distinction can be made between values *missing completely at random* (MCAR), *missing at random* (MAR) and *non-ignorable missing values* (or *informative missing values*). The MCAR variety arise when individuals drop out of the study in a process which is independent of both the observed measurements and those that would have been available had they not been missing; here the observed values effectively constitute a **simple random**

sample of the values for all study participants. Random drop-out (MAR) occurs when the drop-out process depends on the outcomes that have been observed in the past, but, given this information, the drop-out process is conditionally independent of all future (unrecorded) values of the outcome variable following drop-out. Finally, in the case of informative drop-out, the drop-out process depends on the unobserved values of the outcome variable. It is the latter which cause most problems for the analysis of data containing missing values. See also **last observation carried forward**, **attrition** and **imputation**.

Misspecification A term sometimes applied in situations where the wrong model has been assumed for a particular set of observations.

Mixed data Data containing a mixture of continuous variables, ordinal variables and categorical variables.

Mixed effects model A model usually encountered in the analysis of longitudinal data, in which some of the parameters are considered to have fixed effects and some to have random effects. For example, in a clinical trial with two treatment groups in which the response variable is recorded on each subject at a number of visits, the treatments would usually be regarded as having fixed effects and the subjects random effects. [*Journal of Gerontology*, 1996, 51, M86–M91].

MLE Abbreviation for **maximum likelihood estimation**.

MLN A software package for fitting multilevel models.

Mode The most frequently occurring value in a set of observations. Occasionally used as a measure of location. See also **mean** and **median**.

Model A description of the assumed structure of a set of observations, which can range from a fairly imprecise verbal account to, more usually, a formalized mathematical expression of the process assumed to have generated the observed data. The purpose of such a description is to aid in understanding the data. See also **logistic regression**, **multiple regression** and **generalized linear models**.

Model building A procedure which attempts to find the simplest model, for a sample of observations, that provides an adequate fit to the data. See also **parsimony principle**.

Monte Carlo methods Methods for finding solutions to mathematical and statistical problems by simulation. [*Journal of Applied Psychology*, 1998, 83, 164–178].

Moving average A method used primarily for the smoothing of time series, in which each observation is replaced by the average of the observation and its near neighbours. The aim is usually to smooth the series enough to distinguish particular features of interest.

Numerical example

The following data give hourly pain scores for 12 hours for an individual with a migraine headache:

11.6	11.7	5.9	9.8	10.3	12.4
7.4	14.5	15.5	11.5	12.7	10.0

A plot of the original series and a moving average based on sets of four observations is given in Figure 36.

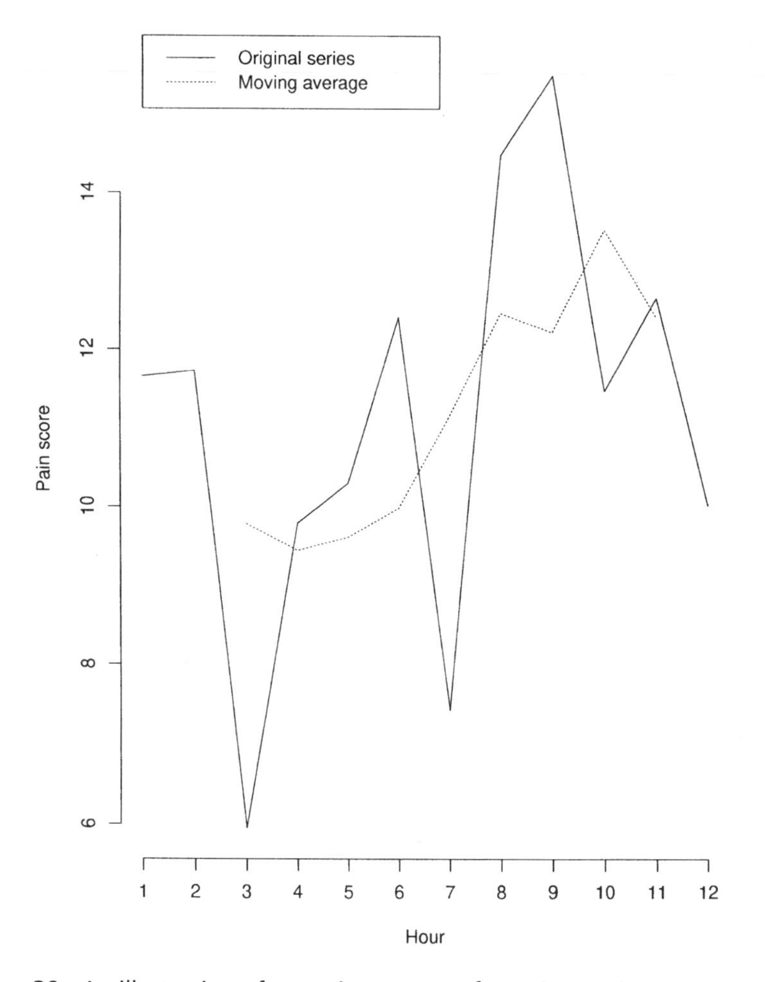

Figure 36 An illustration of a moving average for a time series.

MSE, MSe, MS_e Abbreviations for **mean square**.

Multicentre study A **clinical trial** conducted simultaneously in a number of participating hospitals or clinics, with all centres following an agreed-upon study **protocol** and with independent **random allocation** within each centre. The benefits of such a study include the ability to generalize results to a wider variety of participants and treatment settings than would be possible with a study conducted in a single centre, and the ability to enrol into the study more participants than a single centre could provide.

Multicollinearity A term used in **regression analysis** to indicate situations where the explanatory variables are related by a **linear function**, making the estimation of **regression coefficients** impossible. Including the sum of the explanatory variables in the regression analysis would, for example, lead to this problem. Approximate multicollinearity can also cause problems when estimating regression coefficients. In particular, if the **multiple correlation coefficient** for the regression of a particular explanatory variable on the others is high, then the variance of the corresponding estimated regression coefficient will also be high.

Multidimensional scaling (MDS) A generic term for a class of techniques that attempt to construct a low-dimensional representation of a set of observed **similarity coefficients** or **dissimilarity coefficients**, with the aim of making any structure in the data as clear as possible. See also **individual differences scaling** and **non-metric scaling**.

Mathematical details

The aim is to produce a set of coordinate values to represent the stimuli, so that the distances between them represent the observed proximities. The search is for a low-dimensional space, in which points in the space represent the stimuli, one point representing one stimulus, such that the distances between the points in the space, d_{ij}, match as well as possible, in some sense, the original dissimilarities δ_{ij} or similarities s_{ij}. In a very general sense this simply means that the larger the observed dissimilarity value (or the smaller the similarity value) between two stimuli, the further apart should be the points representing them in the spatial solution.

Numerical example

The data below arise from a famous experiment in which subjects who did not know Morse code listened to pairs of signals (i.e. a sequence of 'dots' and 'dashes') and were required to state whether the two signals they heard were the same or different. Each number in the table is the percentage of a large number of observers who responded 'same' to the row signal followed by the column signal.

	1	2	3	4	5	6	7	8	9	0
1(·----)	84	63	13	8	10	8	19	31	57	55
2(··---)	62	89	54	20	5	14	20	21	16	11
3(···--)	18	64	86	31	23	41	16	17	8	10
4(····-)	5	26	44	89	42	44	32	10	3	3
5(·····)	14	10	30	69	90	42	24	10	6	5
6(-····)	15	14	26	24	17	86	69	14	5	14
7(--···)	22	29	18	15	12	61	85	70	20	13
8(---··)	42	29	16	16	9	30	60	89	61	26
9(----·)	57	39	9	12	4	11	42	56	91	78
0(-----)	50	26	9	11	5	22	17	52	81	94

The two-dimensional solution obtained from applying multidimensional scaling to this table after making it symmetric by taking half the sum of corresponding off-diagonal elements is shown in Figure 37.

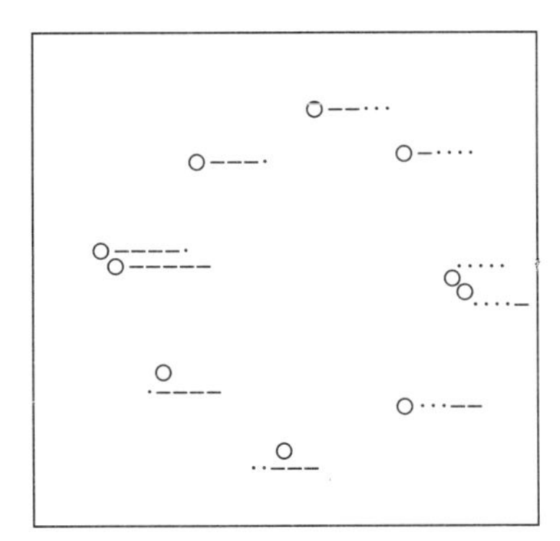

Figure 37 Multidimensional scaling solution for Morse code data.

Multilevel models Models for data that are organized hierarchically – for example, children within families – that allow for the possibility that measurements made on children from the same family are likely to be correlated. See also **mixed effects model**. [*Journal of Applied Psychology*, 1993, 78, 805–814].

Multinomial distribution A generalization of the **binomial distribution** to situations in which r outcomes can occur on each of n trials, where $r > 2$.

Mathematical details

$$\Pr(n_1, n_2, \ldots, n_r) = \frac{n!}{n_1!\, n_2! \cdots n_r!} p_1^{n_1} p_2^{n_2} \cdots p_r^{n_r}$$

where Pr denotes probability and n_i is the number of trials with outcome i, and p_i is the probability of outcome i occurring on a particular trial. The expected value of n_i is np_i and its variance is $np_i(1-p)$. The **covariance** of n_i and n_j is $-np_ip_j$.

Multiple comparison tests Procedures for detailed examination of the differences between a set of means, usually after a general hypothesis that they are all equal has been rejected. No single technique is best in all situations, and a major distinction between techniques is how they control the possible inflation of the Type I error. See also **Bonferroni correction**, **Duncan's multiple range test**, **Scheffé's test**, **post hoc tests** and **Dunnett's test**.

Multiple correlation coefficient The correlation between the observed values of the dependent variable in a multiple regression, and the values predicted by the estimated regression equation. Often used as an indicator of how useful the explanatory variables are in predicting the response. The square of the multiple correlation coefficient gives the proportion of variance of the response variable that is accounted for by the explanatory variables. For a numerical example see **multiple regression**.

Multiple indicator multiple cause model (MIMIC) A structural equation model in which there are multiple indicators and multiple causes of each latent variable.

Multiple regression A model in which a dependent variable is regressed on a number of explanatory variables. Questions of interest usually involve which of the explanatory variables are most useful for determining the response.

Mathematical details

The multiple regression model is that

$$y_i = \beta_0 + \beta_1 x_{1i} + \beta_2 x_{2i} + \cdots + \beta_q x_{qi} + \varepsilon_i$$

where y_i is the dependent variable for subject i, $x_{1i}, \ldots, x_{qi}$ are the values of the q explanatory variables for subject i, and ε_i is a random error term assumed to have a normal distribution with mean zero and variance σ^2. The parameters in the model, $\beta_0, \ldots, \beta_q$, usually known as *regression coefficients*, can be estimated by least squares. An estimated coefficient gives an estimate of the change in the dependent variable produced by a unit change in the corresponding explanatory variable.

Numerical example

Under the Mallow's C_k statistic entry there is a set of data consisting of a dependent variable (income) and three explanatory variables (years of education, years in job and IQ). The estimated regression coefficients and their standard errors for the data are as follows:

Variable	Estimated regression coefficient	SE
IQ	−57.68	235.80
Years of education	3383.91	1293.10
Years in job	1814.39	262.53

The **multiple correlation coefficient** takes the value 0.97. Consequently about 94% (0.97×0.97) of the variance in income is accounted for by the variables IQ, years of education and years in job.

See also **selection methods in regression** and **beta coefficient**. [*Journal of Anxiety Disorders*, 1998, 12, 57–69].

Multiplication rule for probabilities For events A and B that are independent, the probability that both occur is the product of the separate probabilities, i.e.

$$\Pr(A \text{ and } B) = \Pr(A)\Pr(B)$$

where Pr denotes probability. For more than two independent events the rule becomes $\Pr(A \text{ and } B \text{ and } C \text{ and } \ldots) = \Pr(A)\,\Pr(B)\,\Pr(C)\ldots$.

Numerical example

When tossing a fair coin and rolling a fair die, the probability of a head and an even number is

$$\frac{1}{2} \times \frac{1}{3} = \frac{1}{6}$$

Multiplicative model A model in which the combined effect of a number of factors, when applied together, is the product of their separate effects. See also **additive model**.

Multitrait–multimethod model (MTMM) A form of **confirmatory factor analysis** model in which different methods of measurement are used to measure each **latent variable**.

Multivariate analysis A generic term for the many methods of analysis important in investigating **multivariate data**. Examples include **cluster analysis**, **principal components analysis** and **factor analysis**.

Multivariate analysis of variance (MANOVA) A procedure for testing the equality of the **mean vectors** of more than two populations. The technique is directly analogous to the **analysis of variance** of **univariate data**, except that the groups are compared on q response variables simultaneously. In the univariate case, **F-tests** are used to assess the hypotheses of interest. In the multivariate case, however, no single **test statistic** can be constructed that is optimal in all situations.

Mathematical details

Four possible test statistics have been proposed, all based on the three ($q \times q$) matrices:

- **T**, total sums of squares and products matrix
- **W**, within groups sums of squares and products matrix
- **B**, between groups sums of squares and products matrix.

The four test statistics are:

- *Wilks' lambda*:

$$\lambda = \frac{\det(\mathbf{W})}{\det(\mathbf{W}+\mathbf{B})}$$

(where 'det' signifies **determinant**);

- *Roy's largest root criterion*: the largest **eigenvalue** of $\mathbf{BW}^{-1}$;
- the *Hotelling–Lawley trace*: the sum of the eigenvalues of $\mathbf{BW}^{-1}$;
- the *Pillai–Bartlett trace*: the sum of the eigenvalues of $\mathbf{BT}^{-1}$.

It has been found that the differences in **power** between the various test statistics are generally quite small, and so, in most situations, whichever is chosen will not greatly affect conclusions.

Numerical example

These rather ancient data give the ear length measurements of three types of criminals in a UK prison:

Murderers:

Subject	Left ear	Right ear
1	59	59
2	60	65
3	58	62
4	59	59
5	50	48
6	59	65
7	62	62
8	63	62
9	68	72
10	63	66

Criminally insane murderers:

1	70	69
2	69	68
3	65	65
4	62	60
5	59	56
6	55	58
7	60	58
8	58	64
9	65	67
10	67	62

Other criminals:

1	63	63
2	56	57
3	62	62
4	59	58
5	62	58
6	50	57
7	63	63
8	61	62
9	55	59
10	63	63

Applying a MANOVA to these data gives the following results:

	Value	*F*	df	*P*-value
Pillai's trace	0.168	1.241	4, 54	0.30
Wilks' lambda	0.836	1.241	4, 52	0.32
Hotelling's trace	0.190	1.186	4, 50	0.33
Roy's largest root	0.152	2.045	2, 27	0.15

There is no evidence of any group difference.

[*Perceptual and Motor Skills*, 1998, 86, 339–344].

Multivariate data Data for which several measurements or observations are taken on each sample member. For example, measurements on blood pressure, temperature and heart rate for a number of subjects. Such data are usually displayed in the form of a **data matrix**, i.e.

$$\mathbf{X} = \begin{bmatrix} x_{11} & x_{12} & \cdots & x_{1q} \\ x_{21} & x_{22} & \cdots & x_{2q} \\ \vdots & \vdots & \cdots & \vdots \\ x_{n1} & x_{n2} & \cdots & x_{nq} \end{bmatrix}$$

where n is the number of individuals, q the number of variables and x_{ij} the observation on variable j for individual i.

Multivariate distribution The simultaneous **probability distribution** of a set of **random variables**. See also **multivariate normal distribution**.

Multivariate normal distribution The **probability distribution** of a set of variables $\boldsymbol{x}' = [x_1, x_2, \ldots, x_q]$ given by

$$f(x_1, x_2, \ldots, x_q) = (2\pi)^{-q/2}|\boldsymbol{\Sigma}|^{-1/2} \exp -\tfrac{1}{2}(\boldsymbol{x} - \boldsymbol{\mu})'\boldsymbol{\Sigma}^{-1}(\boldsymbol{x} - \boldsymbol{\mu})$$

where μ is the mean vector of the variables and Σ is their variance–covariance matrix. This distribution is assumed by multivariate analysis procedures such as multivariate analysis of variance. See also **bivariate normal distribution**.

Multivariate *t*-test Synonymous with **Hotelling's T^2 test**.

Mutually exclusive events Events that cannot occur jointly, e.g. sleeping and eating.

N

National lotteries Games of chance held to raise money for particular causes. The first held in the UK took place in 1569, principally to raise money for the repair of the Cinque Ports. There were 400 000 tickets or lots, with prizes in the form of plate, tapestries and money. Nowadays lotteries are held in many countries with proceeds either used to augment the Exchequer or to fund good causes. The current UK version began in November 1994 and consists of selecting 6 numbers from 49 for a one pound stake. The winning numbers are drawn 'at random' using one of a number of 'balls-in-drum' type of machine.

Natural pairing See **paired samples**.

Nearest-neighbour clustering Synonym for **single linkage clustering**.

Nearest-neighbour methods Methods of **discriminant analysis** based on studying the individuals most similar to the subject being classified. Classification might then be decided according to a simple majority verdict amongst those most similar of 'nearest' individuals, i.e. a subject would be assigned to the group to which the majority of the 'neighbours' belonged. Simple nearest-neighbour methods consider the single most similar neighbour. More general methods consider the k nearest neighbours, where $k > 1$.

Necessarily empty cells Synonym for **structural zeros**.

Negative binomial distribution In a series of **Bernoulli trials** in which each trial can result in a 'success' or a 'failure', the distribution of the trial number N at which the kth 'success' is achieved.

Mathematical details

The distribution is given by

$$\Pr(N = x) = \binom{x-1}{k-1} p^k (1-p)^{x-k}, \quad x = k, k+1, \ldots$$

where p is the probability of a 'success' on a trial, and Pr denotes probability. The mean of the distribution is k/p and its variance is

$$\frac{k}{p}\left(\frac{1}{p} - 1\right)$$

Numerical example

When tossing a fair coin for which Pr(heads) = Pr(tails) = $\frac{1}{2}$, the probability that the fourth head appears on the 10th trial is

$$\Pr(N = 10) = \binom{10-1}{4-1}\left(\frac{1}{2}\right)^4\left(\frac{1}{2}\right)^6 = 0.082$$

Negative exponential distribution Synonym for **exponential distribution**.

Negative skewness See **skewness**.

Negative synergism See **synergism**.

Nested design A design in which the levels of one or more factors are subsampled within one or more other factors so that, for example, each level of a factor B occurs at only one level of another factor, A. Factor B is said to be nested within factor A. An example might be where interest centres on assessing the effect of clinic and psychologist on a response variable, client satisfaction. The psychologists can only practise at one clinic so they are nested within hospitals. See also **multilevel model**.

Nested model Synonym for **hierarchical model**.

Network In the context of information technology, a linked set of computer systems, capable of sharing computer power and/or storage facilities.

Neural networks See **artificial neural networks**.

Newman–Keuls test A **multiple comparison test** used to investigate in more detail the differences existing between a set of means, as indicated by a significant **F-test** in an **analysis of variance**.

Mathematical details

The test proceeds by arranging the means in increasing order and calculating the **test statistic**

$$S = \frac{\bar{x}_A - \bar{x}_B}{\sqrt{\frac{s^2}{2}\left(\frac{1}{n_A} + \frac{1}{n_B}\right)}}$$

where $\bar{x}_A$ and $\bar{x}_B$ are the two means being compared, s^2 is the **within groups mean square** from the analysis of variance, and n_A and n_B are the numbers of observations in the two groups. Tables of **critical values** of S are available, which depend on a parameter r that specifies the interval between the ranks of the two means being tested. For example, when comparing the largest and smallest of four means, $r = 4$, and when comparing the second smallest and smallest means, $r = 2$.

Numerical example

Suppose in a one-way design there are seven groups with five observations per group and the analysis of variance table gives the following result:

Source	SS	df	MS	F
Groups	45.09	6	7.52	9.40
Error	22.40	28	0.80	
Total	67.49	34		

The ordered group means are:

1	2	3	4	5	6	7
2	2.4	2.6	3.6	4.4	4.8	5

The test for the difference in the smallest and largest mean is

$$S = \frac{5 - 2}{\sqrt{\frac{0.80}{2}\left(\frac{1}{5} + \frac{1}{5}\right)}} = 7.5$$

This is significant when compared with the appropriate critical value from tables. In practice further tests would now be performed. If, however, this test had been non-significant, no additional test would have been performed.

N of 1 clinical trial A special case of a **crossover design** aimed at determining the efficacy of a procedure (or the relative merits of alternative procedures) for a specific individual. The individual is repeatedly given a treatment and placebo, or different treatments, in successive time periods. See also **interrupted time series design**.

Nominal variable Synonym for **categorical variable**.

Non-ignorable missing value Synonymous with **informative missing value**.

Non-linear model A model that is non-linear in the parameters.

Mathematical details

Two examples of non-linear models are

$$y = \beta_1 \, e^{\beta_2 x_1} + \beta_3 \, e^{\beta_4 x_2}$$

$$y = \beta_1 e^{-\beta_2 x}$$

Some such models can be converted into **linear models** by linearization (the second equation above, for example, by taking logarithms throughout). Those that cannot are often referred to as *intrinsically non-linear,* although these can be approximated by linear equations in some circumstances.

Non-metric scaling A form of **multidimensional scaling** in which only the ranks of the observed **dissimilarity coefficients** or **similarity coefficients** are used in producing the required low-dimensional representation of the data. [*Perception and Psychophysics*, 1994, 56, 1–11].

Non-orthogonal designs Synonym for **unbalanced design**.

Non-parametric methods Synonymous with **distribution-free methods**.

Normal approximation A **normal distribution** with mean np and variance $np(1-p)$ that acts as an approximation to a **binomial distribution** as n, the number of trials, increases. The term p represents the probability of a 'success' on any trial.

Normal distribution A **probability distribution** of a **random variable**, x, that is assumed by many statistical methods.

Mathematical details

The distribution is given specifically by

$$f(x) = \frac{1}{\sigma\sqrt{2\pi}} \exp\left[-\frac{1}{2}\frac{(x-\mu)^2}{\sigma^2}\right]$$

μ and σ^2 are, respectively, the mean and variance of x. This distribution is bell shaped. An example is shown in Figure 38.

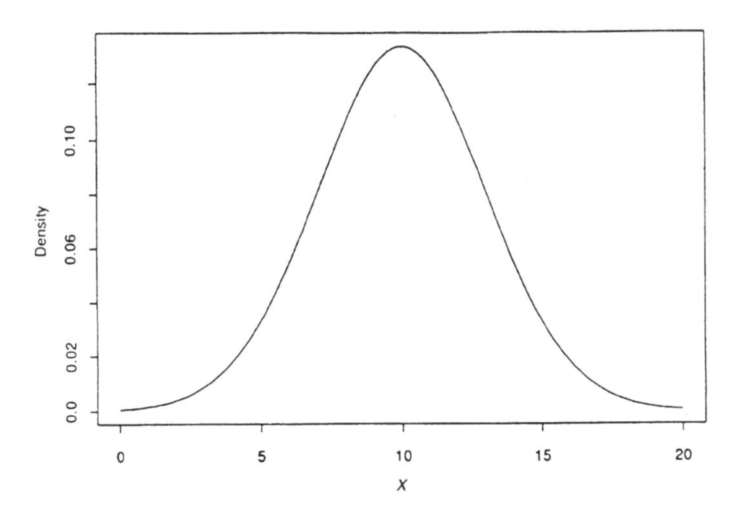

Figure 38 Normal distribution.

Normality A term used to indicate that some variable of interest has a **normal distribution**.

nQuery Advisor A software package useful for determining sample sizes when planning research studies.

Null distribution The **probability distribution** of a **test statistic** when the null hypothesis is true.

Null hypothesis (H_0) The 'no difference' or 'no association' hypothesis to be tested (usually by means of a significance test) against an alternative hypothesis that postulates non-zero difference or association.

Null matrix A matrix in which all elements are zero.

Null vector A **vector**, the elements of which are all zero.

Numerical taxonomy In essence, a synonym for **cluster analysis**.

O

Oblimin A method of factor rotation which allows factors to be correlated.

Oblique factors A term used in factor analysis for common factors that are allowed to be correlated.

Observational study A general term for investigations in which the researcher has little or no control over events, and the relationships between risk factors and outcome measures are studied without the intervention of the investigator. Surveys and most studies in epidemiology fall into this class. See also **experimental study, prospective study** and **retrospective study**.

Occam's razor An early statement of the parsimony principle given by William of Occam (1280–1349), namely 'entia non sunt multiplicanda praeter necessitatem' (a plurality of reasons should not be posited without necessity).

Odds The ratio of the probabilities of the two possible states of a binary variable. See also **odds ratio** and **logistic regression**.

Odds ratio The ratio of the odds for a binary variable in two groups of individuals: for example, males and females. If the two possible states of the variable are labelled 'success' and 'failure', then the odds ratio is a measure of the odds of a success in one group relative to that in the other. When the odds of a success in each group are identical then the odds ratio is equal to unity.

Mathematical details

Suppose data are available in the form:

	Disease present	Disease absent
Risk factor present	a	b
Risk factor absent	c	d

The odds ratio for disease present for the risk factor present and risk factor absent group is

$$\text{odds ratio} = \frac{a/b}{c/d} = \frac{ad}{bc}$$

Numerical example

	Depressed	Not depressed
Chocoholic	70	40
Non-chocoholic	30	60

$$\text{odds ratio} = \frac{70 \times 60}{30 \times 40} = 3.5$$

The odds of being depressed amongst chocoholics is 3.5 times greater than that for non-chocoholics. The odds ratio is of great importance in interpreting the results from a **logistic regression**.

OLS Abbreviation for **ordinary least squares**.

One-sided test A significance test for which the alternative hypothesis is directional: for example, that one population mean is greater than another. The choice between a one-sided and **two-sided test** must be made before any **test statistic** is calculated.

One-tailed test Synonym for **one-sided test**.

One-way design See **analysis of variance**.

Ordered alternative hypothesis A hypothesis that specifies an order for a set of parameters of interest as an alternative to their equality, rather than simply that they are not all equal. For example, in an evaluation of a teaching method given over different periods of time, it might be thought reasonable to postulate that the response variable shows either a monotonic increasing or monotonic decreasing effect with amount of education. In such a case the **null hypothesis** of the equality of, say, a set of m means would be tested against

$$H_1: \ \mu_1 \leq \mu_2 \leq \cdots \leq \mu_m$$

using some suitable test procedure such as **Jonckheere's *k*-sample test**.

Order statistics Particular values in a ranked set of observations. The rth largest value in a sample, for example, is called the rth order statistic.

Ordinal variable A measurement that allows a sample of individuals to be ranked with respect to some characteristic, but where differences at different points of the scale are not necessarily equivalent. For example, anxiety might be rated on a scale 'none', 'mild', 'moderate' and 'severe', with the values 0, 1, 2, 3, being used to label the categories. A patient with anxiety score of 1 could be ranked as less anxious than one given a score of 3, but patients with scores 0 and 2 do not necessarily have the same difference in anxiety as patients with scores 1 and 3. See also **categorical variable** and **continuous variable**.

Ordinary least squares (OLS) See **least squares estimation**.

Ordination The process of reducing the dimensionality (i.e. the number of variables) of multivariate data by deriving a small number of new variables that contain much of the information in the original data. The reduced data set is often more useful for investigating possible structure in the observations. See also **principal components analysis** and **multidimensional scaling**.

Orthogonal matrix A square matrix which on multiplication by its transpose results in an identity matrix. Arises in accounts of many methods of multivariate analysis.

Orthonormal contrasts See **orthogonal contrasts**.

Outcome variable Synonym for **dependent variable**.

Outlier An observation that appears to deviate markedly from the other members of the sample in which it occurs. In the set of exam scores 47, 52, 41, 55, 98, for example, 98 might be considered an outlier. More formally the term refers to an observation which appears to be inconsistent with the rest of the data, relative to an assumed model. Such extreme observations may reflect some abnormality in the measured characteristic of a subject, or they may result from an error in the measurement or recording. See also **outside observation**.

Outside observation An observation falling outside the limits

$$F_{\mathrm{L}} - 1.5(F_{\mathrm{U}} - F_{\mathrm{L}}), \qquad F_{\mathrm{U}} + 1.5(F_{\mathrm{U}} - F_{\mathrm{L}})$$

where F_{U} and F_{L} are the upper and lower quartiles of a sample. Such observations are usually regarded as being extreme enough to be potential outliers. See also **box-and-whisker plot**.

Overfitted models Models that contain more unknown parameters than can be justified by the data.

Overidentified model See **identification**.

Overmatching A term applied to studies involving matching when the matching variable is strongly related to exposure but not to disease risk.

Overparameterized model A model with more parameters than observations for estimation.

Mathematical details

The following simple model for a one-way design in analysis of variance:

$$y_{ij} = \mu + \alpha_i + \varepsilon_{ij} \qquad (i = 1, 2, \ldots, g, \quad j = 1, 2, \ldots, n_i)$$

(where g is the number of groups, n_i the number of observations in group i, y_{ij} represents the jth observation in the ith group, μ is the grand mean effect and α_i the effect of group i) has $g + 1$ parameters but only g group means to be fitted. It is overparameterized unless some constraints are placed on the parameters, for example, that

$$\sum_{i=1}^{g} \alpha_i = 0$$

See also **identification**.

Overviews Synonym for **meta-analysis**.

P

Paired samples Two samples of observations with the characteristic feature that each observation in one sample has one and only one matching observation in the other sample. There are several ways in which such samples can arise in psychological investigations. The first, *self-pairing*, occurs when each participant serves as his or her own control, as in, for example, therapeutic trials in which each participant receives both treatments, one on each of two separate occasions. Next, *natural pairing* can arise, particularly, for example, in laboratory experiments involving litter-mate controls. Lastly, *artificial pairing* may be used by an investigator to match the two participants in a pair on important characteristics likely to be related to the response variable.

Paired samples *t*-test Synonym for **matched pairs *t*-test**.

Panel study A study in which a group of people, the 'panel', are interviewed or surveyed with respect to some topic of interest on more than one occasion. Essentially equivalent to a longitudinal study, although there may be many response variables observed at each time point.

Parallel groups design A simple experimental set-up in which two different groups, for example, are studied concurrently.

Parameter A numerical characteristic of a population or a model. The probability of a 'success' in a binomial distribution, for example, or the mean of a normal distribution.

Parametric hypothesis A hypothesis concerning the parameter(s) of a distribution. For example, the hypothesis that the mean of a population equals the mean of a second population, when the populations are each assumed to have a normal distribution.

Parametric methods Procedures for testing hypotheses about parameters in a population described by a specified distributional form, often a normal distribution. Student's *t*-test is an example of such a method. See also **distribution-free methods**.

Parsimony principle The general principle that, amongst competing models, all of which provide an adequate fit for a set of data, the one with the fewest parameters is to be preferred. See also **Occam's razor**.

Partial correlation The correlation between a pair of variables after adjusting for the effect of a third.

Mathematical details

Can be calculated from the sample **correlation coefficients** of each of the three pairs of variables involved as

$$r_{12.3} = \frac{r_{12} - r_{13}r_{23}}{\sqrt{(1 - r_{13}^2)(1 - r_{23}^2)}}$$

Numerical example

The correlations between weight (1), scores on a test for arithmetic (2) and chronological age (3) were found to be

$$r_{12} = 0.50, \qquad r_{13} = 0.80, \qquad r_{23} = 0.60$$

The partial correlation between weight and the arithmetic scores with the effect of chronological age partialled out is

$$r_{12.3} = \frac{0.50 - (0.80)(0.60)}{\sqrt{(1 - 0.80^2)(1 - 0.60^2)}} = 0.04$$

The partial correlation is negligible, indicating that the correlation between weight and arithmetic scores arises only because of the correlation of each with age.

Path analysis A tool for evaluating the interrelationships among variables by analysing their correlational structure. The relationships between the variables are often illustrated graphically by means of a *path diagram*, in which single-headed arrows indicate the direct influence of one variable on another, and curved double-headed arrows indicate correlated variables. Originally introduced for simple regression models for observed variables, the method has now become the basis for more sophisticated procedures such as **confirmatory factor analysis** and **structural equation modelling**, involving both **manifest variables** and **latent variables**. An example of a path diagram representing a correlated two-factor model is shown in Figure 39. [*Perceptual and Motor Skills*, 1998, 86, 139–145].

Path coefficient Synonym for **standardized regression coefficient**.

Path diagram See **path analysis**.

Pearson's chi-squared statistic See **chi-squared statistic**.

Pearson's product moment correlation coefficient See **correlation coefficient**.

Pearson's residual A model diagnostic used particularly in the analysis of **contingency tables** and **logistic regression**.

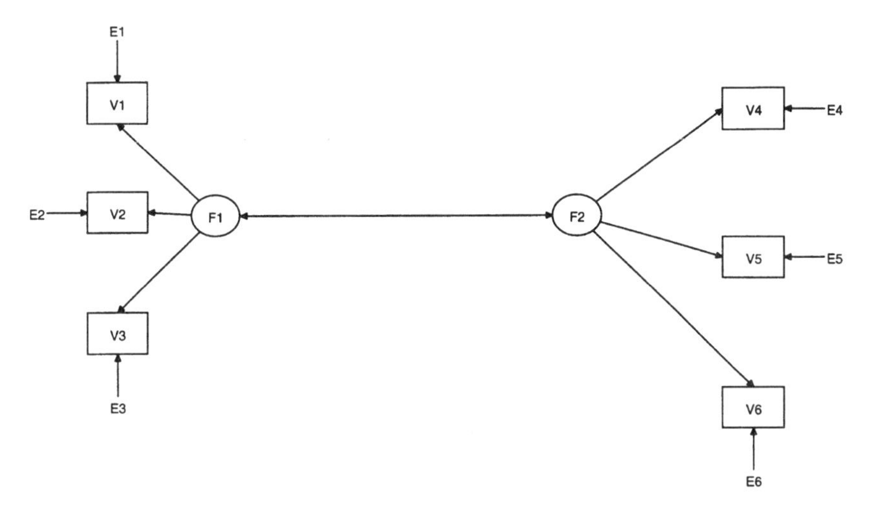

Figure 39 A path diagram for a correlated two-factor model.

Mathematical details

$$r = \frac{O - E}{\sqrt{E}}$$

where O represents the observed value and E the corresponding predicted value under some model. Such residuals, if the assumed model is true, have approximately a **standard normal distribution**, so that values of r outside the range -2.0, 2.0 suggest aspects of the current model that are inadequate.

Numerical example

The following 2×2 contingency table gives the incidence of suicidal feelings in samples of psychotic and neurotic patients:

	Psychotic	Neurotic	Total
Suicidal feelings	2 (4)	6 (4)	8
No suicidal feelings	18 (16)	14 (16)	32
Total	20	20	

The values in parentheses give the estimated expected values under the hypothesis that suicidal feelings are independent of type of patient. Pearson's residuals for the 4 cells are

$$\begin{matrix} -1 & 1 \\ 0.5 & -0.5 \end{matrix}$$

These imply that independence is a reasonable assumption.

Percentile The set of divisions that produce exactly 100 equal parts in a series of continuous values, such as blood pressure, weight and height. Thus, a person with an IQ above the 80th percentile has a greater IQ value than over 80% of the other recorded values.

Per-comparison error rate The significance level at which each test or comparison is carried out in an experiment. See also **per-experiment error rate**.

Per-experiment error rate The probability of incorrectly rejecting at least one null hypothesis in an experiment involving one or more tests or comparisons, when the corresponding null hypothesis is true in each case. See also **per-comparison error rate**.

Perspective plot See **contour plot**.

Phi-coefficient A measure of association of the two variables forming a **two-by-two contingency table**.

Mathematical details

$$\phi = \sqrt{\frac{\chi^2}{N}}$$

where χ^2 is the usual **chi-squared statistic** for the independence of the two variables and N is the sample size. The coefficient has a maximum value of unity, and the closer its value to unity, the stronger the association between the two variables.

Numerical example

For the 2×2 table given under the **Pearson's residual** entry the phi-coefficient takes the value

$$\sqrt{2.5/40} = 0.25$$

See also **Cramér's V** and **contingency coefficient**.

Pie chart A widely used graphical technique for presenting the distributions associated with the observed values of a categorical variable. The chart consists of a circle sub-divided into sectors whose sizes are proportional to the quantities (usually percentages) they represent. An example involving perception of health is given in Figure 40. Such displays are popular in the media but have little relevance for serious scientific work, when other graphics are generally far more useful. See also **bar chart** and **dot plot**.

Pilot study A small-scale investigation carried out before the main survey, primarily to gain information and to identify problems relevant to the survey proper.

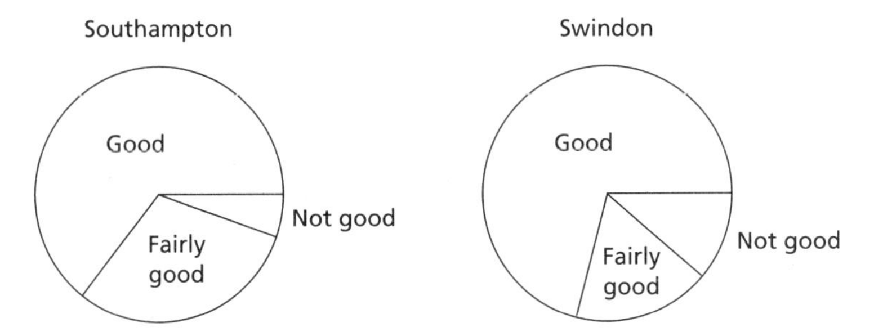

Figure 40 Pie charts for subjective assessment of health.

Placebo A treatment designed to appear exactly like a comparison treatment, but which is devoid of the active component.

Placebo effect A well-known phenomenon in psychology and medicine in which participants in a study given only inert substances often show subsequent clinical improvement in their symptoms when compared with participants not so 'treated'.

Placebo reactor A term sometimes used for those participants in a clinical trial who report side effects normally associated with the active treatment while receiving a placebo.

Planned comparison A comparison between a set of means suggested before data are collected. Usually more powerful than a general test for mean differences. See also **post hoc comparison**.

Platykurtic curve See **kurtosis**.

Point-biserial correlation A special case of Pearson's product moment correlation coefficient used when one variable is continuous (y) and the other is a binary variable (x) representing a natural dichotomy.

Mathematical details

$$r_{\text{pb}} = \frac{\bar{y}_1 - \bar{y}_0}{s_y}\sqrt{p\,q}$$

where $\bar{y}_1$ is the sample mean of the y variable for those individuals with $x = 1$, $\bar{y}_0$ is the sample mean of the y variable for those individuals with $x = 0$, s_y is the standard deviation of the y values, p is the proportion of individuals with $x = 1$, and $q = 1 - p$ is the proportion of individuals with $x = 0$.

Numerical example

The data below show IQ scores for a number of boys and girls of the same age:

Boys	101	107	98	105	110	99	93	112			
Girls	93	94	108	109	109	91	115	112	98	99	102

The point-biserial correlation in this case is $r_{\mathrm{pb}} = 0.006$, indicating that gender and IQ are not associated.

See also **biserial correlation**.

Poisson distribution The **probability distribution** of the number of occurrences, N, of some random event, in an interval of time or space.

Mathematical details

The distribution is given explicitly

$$\Pr(N = n) = \frac{e^{-\lambda}\lambda^n}{n!}, \qquad n = 0, 1, \ldots$$

The mean and variance of a variable with such a distribution are both equal to λ.

Poisson regression A method for the analysis of the relationship between an observed count with a **Poisson distribution** and a set of explanatory variables.

Polychotomous variables Strictly, variables that can take more than two possible values, but since this would include all but **binary variables**, the term is conventionally used for **categorical variables** with more than two categories.

Polynomial regression A **linear model** in which powers and possibly cross-products of explanatory variables are included. For example

$$y = \beta_0 + \beta_1 x + \beta_2 x^2$$

Population In statistics this term is used for any finite or infinite collection of 'units', which are often people, but may be, for example, institutions, events, etc. See also **sample**.

Positive skewness See **skewness**.

Positive synergism See **synergism**.

Posterior distribution A **probability distribution** that summarizes information about a **random variable** or parameter after, or *a posteriori* to, having obtained new information from empirical data. Used almost entirely within the context of **Bayesian inference**. See also **prior distribution**.

Post hoc comparison An analysis most explicitly planned at the start of a study but suggested by an examination of the data. See also **multiple comparison procedure**, **data dredging** and **planned comparisons**.

Power The probability of rejecting the null hypothesis when it is false. Power gives a method of discriminating between competing tests of the same hypothesis, the test with the higher power being preferred. It is also the basis of procedures for estimating the sample size needed to detect an effect of a particular magnitude.

Numerical example

As an example it can be shown that in a study in which an investigator wishes to detect a difference between **incidences** of 0.80 and 0.50 with a power of 95% when testing at the 5% level, 49 participants are needed in each group.

See also **nQuery Advisor**.

Precision A term applied to the likely spread of estimates of a parameter in a statistical model. Measured by the **standard error** of the estimator; this can be decreased, and hence precision increased, by using a larger sample size.

Prevalence The number of persons in a defined population who have a disease or condition at some given point in time. Also often used for the proportion of individuals with disease in the population, although this is more properly termed the *prevalence rate*. See also **incidence**.

Prevalence rate See **prevalence**.

Principal components analysis A procedure for analysing **multivariate data**, which transforms the original variables into new ones that are uncorrelated with each other, and each of which accounts for a smaller proportion of the variance in the data than does the preceding (new) variable. The aim of the method is to reduce the dimensionality of the data. The new variables, the principal components, are defined as **linear functions** of the original variables. If the first few principal components account for a large percentage of the variance of the observations (say above 70%), they can be used both to simplify subsequent analyses and to display and summarize the data in a parsimonious manner.

Mathematical details

The first principal component of the observations is that linear combination, y_1, of the original variables

$$y_1 = \alpha_{11} x_1 + \alpha_{12} x_2 + \cdots + a_{1q} x_q$$

whose sample variance is greatest for all coefficients $\alpha_{11}, \ldots, \alpha_{1q}$.

The second principal component is the linear combination of the original variables having greatest variance, subject to being uncorrelated with y_1.

Subsequent principal components are defined in similar fashion, each being specified to be uncorrelated with preceding components.

The required coefficients are given by the **eigenvectors** of the **variance–covariance matrix** or **correlation matrix** of the observations. The corresponding **eigenvalues** give the variances of the components.

Example

Suppose we have a correlation matrix for two variables

$$R = \begin{bmatrix} 1.0 & r \\ r & 1.0 \end{bmatrix}$$

In this case the two principal components can be shown to have the form

$$y_1 = \frac{1}{\sqrt{2}}(x_1 + x_2), \quad \text{var}(y_1) = 1 + r$$

$$y_2 = \frac{1}{\sqrt{2}}(x_1 - x_2), \quad \text{var}(y_2) = 1 - r$$

If $r < 0$ the order of the eigenvalues, and hence of the principal components, is reversed.

See also **factor analysis, Kaiser's rule** and **scree plot**. [*Schizophrenia Research*, 1996, 20, 221–229].

Prior distribution A **probability distribution** that summarizes information about a **random variable** or parameter known or assumed, prior to obtaining further information from empirical data. Used almost entirely within the context of **Bayesian inference**.

Probability A measure associated with an event A and denoted by $\Pr(A)$ which takes a value such that $0 \leq \Pr(A) \leq 1$. Essentially the quantitative expression of the chance that an event will occur. In general the higher the value of $\Pr(A)$ the more likely it is that the event will occur. If an event cannot happen, $\Pr(A) = 0$; if an event is certain to happen, $\Pr(A) = 1$. Numerical values can be assigned in simple cases by one of the following two methods:

1. If the **sample space** can be divided into subsets of n ($n \geq 2$) equally likely outcomes and the event A is associated with r ($0 \leq r \leq n$) of these, then $\Pr(A) = r/n$.
2. If an experiment can be repeated a large number of times, n, and in r cases the even A occurs, then r/n is called the relative frequency of A. If this leads to a limit as $n \to \infty$, this limit is $\Pr(A)$.

See also **addition rule for probabilities** and **multiplication rule for probabilities**.

Probability density See **probability distribution**.

Probability distribution For a discrete random variable, a mathematical formula that gives the probability of each value of the variable. See, for example, **binomial distribution** and **Poisson distribution**. For a continuous random variable, a curve described by a mathematical formula which specifies, by way of areas under the curve, the probability that the variable falls within a particular interval. Examples include the **normal distribution** and the **exponential distribution**. In both cases the term *probability density* may also be used. (A distinction is sometimes made between 'density' and 'distribution', when the latter is reserved for the probability that the random variable falls below some value. In this dictionary, however, the latter is termed the *cumulative probability distribution*, and probability distribution and probability density are used synonymously.)

Product limit estimator A procedure for estimating the survival function for a set of survival times, some of which may be censored observations. The idea behind the procedure involves the product of a number of conditional probabilities, so that, for example, the probability of a patient surviving two days after a liver transplant can be calculated as the probability of surviving one day, multiplied by the probability of surviving the second day given that the patient survived the first day.

Mathematical details

The estimator, $\hat{S}(t)$, is given specifically as

$$\hat{S}(t) = \prod_{j|t(w)^{i}t} \left(1 - \frac{d_j}{r_j}\right)$$

where $\hat{S}(t)$ is the estimated survival function at time t, $t_{(1)} \leq t_{(2)} \leq \cdots \leq t_{(n)}$ are the ordered survival times, r_j is the number of variables at risk at time $t_{(j)}$, and d_j is the number of individuals who experience the event of interest at time $t_{(j)}$. (Individuals censored at time $t_{(j)}$are included in r_j). The resulting estimates form a step function that can be plotted to give a graphical display of survival experience. An example appears in Figure 41.

[*Journal of Mental Health UK*, 1998, 7, 49–57].

Profile analysis A term sometimes used for the **analysis of variance** of **longitudinal data**.

Programming The act of planning and producing a set of instructions to solve a problem by computer. See also **algorithm**.

Prospective study Studies in which individuals are followed up over a period of time. A common example of this type of investigation is where samples of individuals exposed and not exposed to a possible **risk factor** for a particular disease are followed forward in time, to determine what happens to them with respect to the illness under investigation. At the end of a suitable time period a comparison of the **incidence** of the disease amongst the exposed and non-exposed is made. A classic example of such a study is that undertaken amongst British doctors in the 1950s, to investigate whether any relationship existed between smoking and death from lung cancer. See also **longitudinal study**, **retrospective study** and **cohort study**.

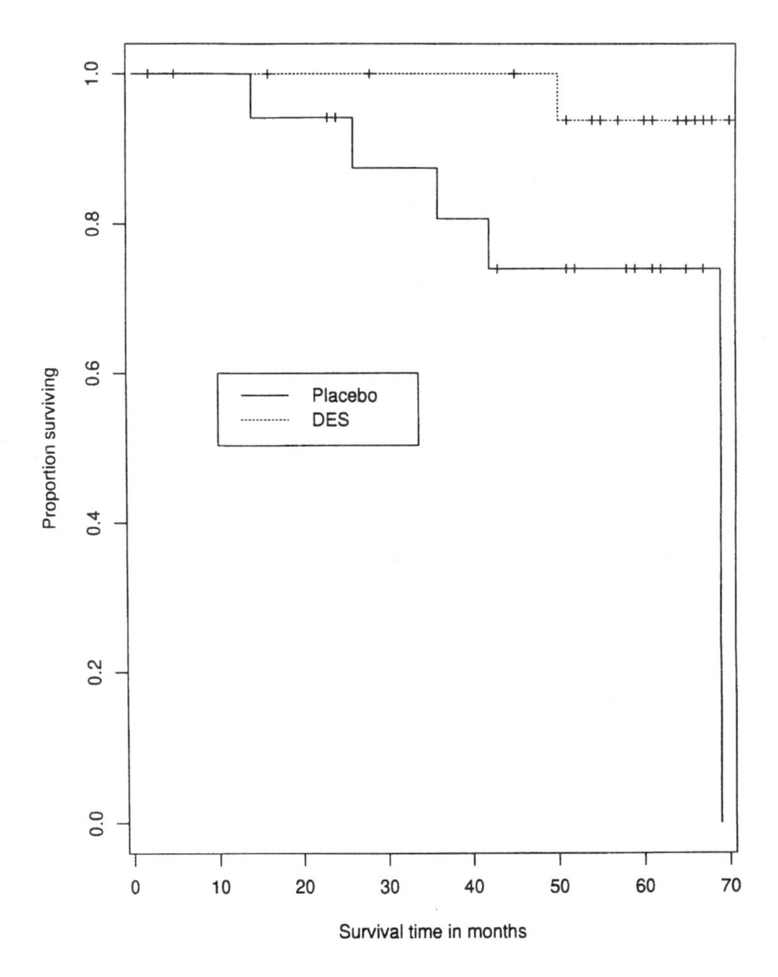

Figure 41 Survival curves found by application of product limit estimator, in a clinical trial of two treatments for cancer.

Protocol A formal document outlining the proposed procedures for carrying out a research investigation, in particular, a clinical trial. The main features of the document are study objectives, participant selection criteria, treatment schedules, methods of patient evaluation, trial design, procedures for dealing with protocol violators and plans for statistical analysis.

Protocol violators Participants who either deliberately or accidentally have not followed one or other aspect of the protocol for carrying out a research study. For example, they may not have taken their prescribed medication.

Proximity matrix A term used to refer to either a similarity matrix or a dissimilarity matrix.

Pseudorandom numbers A sequence of numbers generated by a specific computer algorithm which satisfy particular statistical tests of randomness. So although not random, the numbers appear so.

Psychlit CDs containing literature on psychology and material relevant to psychology. One database covers 30 languages from over 45 countries. The other covers English language books and book chapters. The publishers are the American Psychological Association.

Publication bias Bias that occurs because journal editors are more likely to accept a paper if a significant effect has been demonstrated. See also **funnel plot** and **meta-analysis**.

***P*-value** The probability of the observed data (or data showing a more extreme departure from the null hypothesis) conditional on the null hypothesis being true. See also **misinterpretation of *P*-values**, **significance test** and **significance level**.

Q-techniques Data analysis methods that look for relationships between the individuals in a set of multivariate data. Includes cluster analysis and multidimensional scaling, although the term is most commonly used for a type of factor analysis applied to an $n \times n$ matrix of 'correlations' between individuals rather than between variables. See also **R-techniques**.

Quadratic discriminant analysis See **discriminant analysis**.

Quantiles Divisions of a probability distribution or frequency distribution into equal, ordered subgroups: for example, quartiles or percentiles.

Quartiles The values that divide a frequency distribution or probability distribution into four equal parts.

Quasi-experiment A term used for studies that resemble experiments but are weak on some of the characteristics, particularly that manipulation of participants to groups is not under the investigator's control. For example, if interest centred on the psychological health effects of a natural disaster, those who experience the disaster can be compared with those who do not, but participants cannot be deliberately assigned (randomly or not) to the two groups. See also **prospective studies, experimental design** and **clinical trials**.

Quasi-independence A form of independence for a contingency table, conditional on restricting attention to a particular part of the table only. For example, in the following table showing the social class of sons and their fathers, it might be of interest to assess whether, once a son has moved out of his father's class, his destination class is independent of that of his father. This would entail testing whether independence holds in the table after ignoring the entries in the main diagonal.

	Son's social class		
Father's social class	**Upper**	**Middle**	**Lower**
Upper	588	395	159
Middle	349	714	447
Lower	114	320	411

Quick and dirty methods A term once applied to many distribution-free methods, presumably to highlight their general ease of computation and their imagined inferiority to the corresponding parametric procedure.

R

Random allocation A method for forming experimental and control groups. Participants receive the experimental or control condition on the basis of the outcome of a chance event: for example, tossing a coin. The method provides an impartial procedure for allocation of treatments to individuals, free from personal biases, and ensures a firm footing for the application of significance tests and most of the rest of the statistical methods likely to be used. Additionally, the method distributes the effects of concomitant variables, both observed and unobserved, in a statistically acceptable fashion.

Random effects The effects attributable to an (usually) infinite set of levels of a factor, of which only a **random sample** occurs in the data. For example, the investigator may be interested in the effects of a particular class of drug on behaviour, and uses a random sample of drugs from the class in a study. A *random-effects model* arises when all the factors in the investigation are of this kind. See also **fixed effects**.

Random-effects model See **random effects**.

Random events Events which do not have deterministic regularity (observations of them do not necessarily yield the same outcome) but with some degree of statistical regularity (indicated by the statistical stability of their frequency).

Randomization tests Procedures for determining statistical significance directly from data, without recourse to some particular **sampling distribution**. The data are divided (permuted) repeatedly between treatments, and for each division (permutation) the relevant **test statistic** (for example t or F) is calculated to determine the proportion of the data permutations that provide as large a test statistic as that associated with the observed data. If that proportion is smaller than some significance level, α, the results are significant at the α level. See also **bootstrap** and **jackknife**.

Randomized block design (RBD) An **experimental design** in which the treatments in each **block** are assigned to the experimental units in random order.

Randomized clinical trial (RCT) Synonymous with **clinical trial**.

Random model Governed by chance; not completely determined by other factors. Non-deterministic.

Random sample Either a set of n independent and identically distributed **random variables**, or a sample of n individuals selected from a population in such a way that each sample of the same size is equally likely.

Random variable A variable, the values of which occur according to some specified **probability distribution**.

Random variation The variation of data unexplained by identifiable sources.

Range The difference between the largest and smallest observations in a data set. Often used as an easy-to-calculate measure of the **dispersion** in a set of observations – but not recommended for this task, because of its sensitivity to **outliers**.

Rank correlation coefficients **Correlation coefficients** that depend only on the **ranks** of the variables, not on their observed values. Examples include **Kendall's tau** and **Spearman's rho**.

Ranking The sorting of a set of variable values into ascending or descending order.

Ranks The relative positions of the members of a sample with respect to some characteristic.

Rasch model A model for tests of ability originally proposed in the context of oral reading tests. By considering a number of tests and individuals, **maximum likelihood estimation** can be used to find estimates of parameters in the model representing each individual's ability and each test's difficulty reference. [*Journal of Child Psychology and Psychiatry and Allied Disciplines*, 1998, 39, 215–224].

Ratio variable A continuous variable ratio that has a fixed rather than an arbitrary zero point. Examples are height, weight and temperature measured in degrees Kelvin. See also **categorical variable** and **ordinal variable**.

RBD Abbreviation for **randomized block design**.

RCT Abbreviation for **randomized clinical trial**.

Recall bias A possible source of **bias**, particularly in a **retrospective study**, caused by differential recall amongst cases and controls, in general by under-reporting of exposure in the control group. In a study comparing cases with depression, for example, and a non-depressed control group, the cases may remember more distressing life events than the controls. See also **ascertainment bias**.

Receiver operating characteristic (ROC) curves A plot of the **sensitivity** of a diagnostic test against one minus its **specificity**, as the cut-off criterion for indicating a positive test is varied. Often used in choosing between competing tests, although it takes no account of the **prevalence** of the disease being tested for.

Numerical example

Consider the following ratings from 1 (definitely well) to 5 (definitely ill) arising from 50 well and 50 ill subjects.

	1	2	3	4	5	Total
Well	4	17	20	8	1	50
Ill	3	3	17	19	8	50

If the rating of 5 is used as the cut-off for identifying cases that are ill then the sensitivity is estimated as $8/50 = 0.16$, and the specificity as $49/50 = 0.98$. Now using the rating 4 as cut-off leads to a sensitivity of $27/50 = 0.54$ and a specificity of $41/50 = 0.82$. The values of (sensitivity, 1 − specificity), as the cut-off decreases from 5 to 1, are (0.16, 0.02), (0.54, 0.18), (0.88, 0.58), (0.94, 0.92), (1.00, 1.00). These points are plotted in Figure 42. This is the required receiver operating characteristic curve.

[*Psychological Assessment*, 1997, 9, 171–176].

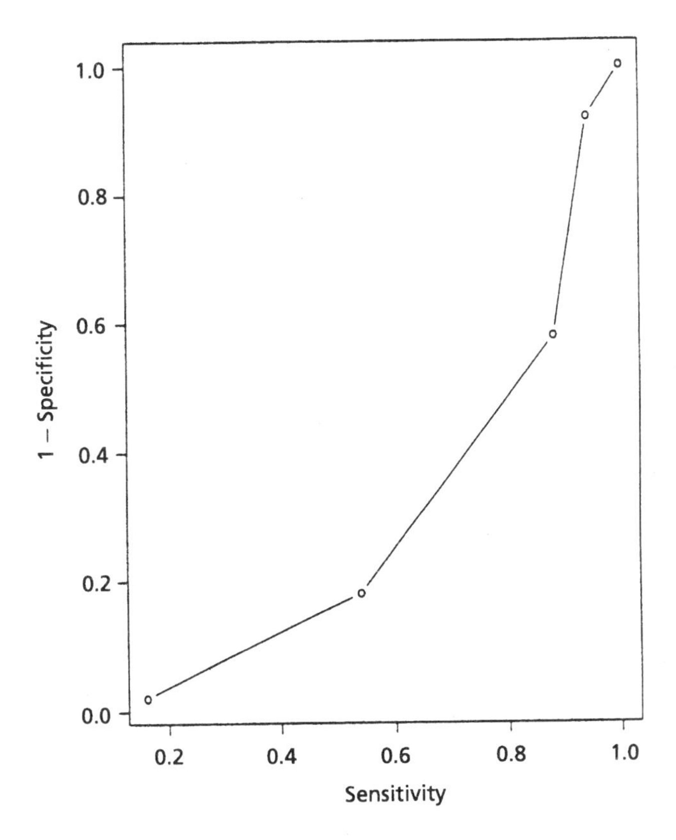

Figure 42 Receiver operating characteristics.

Reciprocal transformation A transformation of the form $y = 1/x$, which is particularly useful for certain types of variables since it may improve their distributional properties for some types of analysis, whilst retaining a direct interpretation for the transformed variable. Resistances, for example, become conductances, and reaction times become speeds.

Regression analysis A frequently applied statistical technique that serves as the basis for studying and characterizing a system of interest, by formulating a reasonable mathematical model of the relationship between a response variable and a set of explanatory variables. **Multiple regression** and **logistic regression** are the most commonly used types.

Regression coefficients See **multiple regression**.

Regression diagnostics Procedures designed to investigate the assumptions underlying a **regression analysis** – for example, normality, homogeneity of variance – or to examine the **influence** of particular data points or small groups of data points on the estimated **regression coefficients**. An essential part of all regression analyses. See also **residuals**.

Regression through the origin In some situations a relationship between two variables estimated by **regression analysis** is expected to pass through the origin, because the true mean of the dependent variable is known to be zero when the value of the explanatory variable is zero. In such situations the **linear regression** model is forced to pass through the origin by setting the intercept parameter to zero and estimating only the slope parameter.

Regression to the mean The process first noted by Sir Francis Galton that 'each peculiarity in man is shared by his kinsmen, but on the average to a less degree'. Hence the tendency, for example, for tall parents to produce tall off-spring but who, on average, are shorter than their parents. The term is now generally used to label the phenomenon that a variable that is extreme on its first measurement will tend to be closer to the centre of the distribution for a later measurement. For example, in a screening programme for anxiety, only persons with high anxiety are asked to return for a second measure. On the average, the second measure taken will be less than the first. [*Journal of Family Psychology*, 1997, 11, 351–360].

Reification The process of naming **latent variables** and the consequent discussion of such things as quality of life and racial prejudice as though they were physical quantities in the same sense as, for example, length and weight.

Relative risk A measure of the association between exposure to a particular factor and risk of a certain outcome, calculated as

$$\text{relative risk} = \frac{\text{incidence rate among exposed}}{\text{incidence rate among non-exposed}}$$

Thus a relative risk of 5, for example, means that an exposed person is five times as likely to have the disease than one who is not exposed. Relative risk does *not* measure the probability that someone with the factor will develop the disease. The disease may be rare among both the non-exposed and the exposed. See also **incidence** and **attributable risk**. [*Work and Stress*, 1997, 11, 103–117].

Reliability The extent to which the same measurements of individuals obtained under different conditions yield similar results. See also **intra-class correlation** and **kappa coefficient**.

Reliable change indicator A measure of the value of the clinical change following an intervention that is often more important to a psychologist than a statistically significant change. The calculation of the indicator often takes into account the reliability of the measuring instruments and/or **regression to the mean**. Several methods of calculation exist in the literature which depend on different aims and assumptions. The three different goals of an intervention can be exemplified by: (a) the patient should now lie outside the dysfunctional population range (i.e. two standard deviations above the mean for the dysfunctional group in the direction of higher functioning), (b) the patient's scores now lie within the functional rather than the dysfunctional range, or (c) the patient's scores are now closer to the mean for the functional population than the dysfunctional population. [*Journal of Consulting and Clinical Psychology*, 1991, 59, 12–19].

Repeated-measures data See **longitudinal data**.

Replicate observation An independent observation obtained under conditions as close to the original as the nature of the investigation will permit.

Residual The difference between the observed value of a dependent variable (y_i) and the value predicted by some model of interest ($\hat{y}_i$). Examination of a set of residuals, usually by informal graphical techniques, allows the assumptions made in the model-fitting exercise – for example, normality, homogeneity of variance – to be checked. Observations with 'large' residuals may give cause for concern. Residuals may need to be standardized to make them of practical use.

Mathematical details

A basic residual r_i is defined as

$$r_i = y_i - \hat{y}_i$$

A standardized version is given by

$$r_i^{(s)} = \frac{y_i - \hat{y}_i}{\sqrt{1 - h_i}}$$

where s^2 is the estimated residual variance after fitting the model of interest, and h_i is the ith diagonal element of the **hat matrix**.

An alternative definition of a standardized residual (sometimes known as the *Studentized residual*), is

$$r_i^{(st)} = \frac{y_i - \hat{y}_i}{s_{(-i)}\sqrt{1 - h_i}}$$

where $s_{(-i)}^2$ is the estimated residual variance calculated from the sample values minus observation i.

Numerical example

Under the **multiple regression** entry there is a set of data involving a dependent variable and three explanatory variables. The various types of residuals from fitting a multiple regression model are as follows:

Individual	Observed value	Predicted value	r_i	$r_i^{(s)}$	$r_i^{(st)}$
1	8 000	9291.2	−1291.2	−0.65	−0.62
2	10 000	10 341.6	−341.6	−0.19	−0.17
3	10 500	7837.0	2663.0	1.44	1.62
4	15 000	15 828.3	−828.3	−0.88	−0.86
5	12 000	14 979.2	−2979.2	−1.71	−2.26
6	25 000	23 790.1	1209.9	0.65	0.62
7	17 000	17 813.7	−813.7	−0.65	−0.61
8	30 000	27 953.0	2047.8	1.31	1.41
9	6 000	4439.0	1561.0	0.90	0.88
10	20 000	21 227.9	−1227.9	−0.64	−0.60

The standardized residual and the Studentized residual can be compared with −2 and 2 to identify possibly problem observations. Here individual 5 has an observed salary well below the predicted value.

See also **regression diagnostics**.

Residual sum of squares See **analysis of variance**.

Response feature analysis An approach to the analysis of **longitudinal data** involving the calculation of suitable summary measures from the set of repeated measures on each participant. For example, the mean of the participant's measurements might be calculated or the maximum value of the response variable over the repeated measurements, etc. Simple methods such as **Student's t-tests** or **Mann–Whitney tests** are then applied to these summary measures to assess differences between conditions.

Numerical example

In a clinical trial in which women who had suffered an episode of post-natal depression were randomized to receive either an oestrogen transdermal patch or a placebo patch, a composite measure of depression was taken at 2-monthly intervals post-randomization for a year. Part of the data is shown below:

		Visit						
	Subject	1	2	3	4	5	6	Mean
Active treatment	1	27	20	15	15	10	9	16.00
	2	14	12	15	12	9	6	11.33
	3	19	9	9	12	5	7	10.17
	4	16	13	11	11	11	11	12.17
	5	15	18	12	9	8	6	11.33
Placebo	1	26	23	18	17	12	10	17.67
	2	20	19	11	9	8	6	12.17
	3	28	26	27	27	25	20	25.50
	4	20	15	20	17	15	14	16.83
	5	13	12	9	9	8	7	9.67

Using the mean as the chosen summary measure for each woman's six observations, we can compare the two treatment groups by a Student's t-test based on the mean values.

- Active treatment: mean = 12.20; variance – 5.02
- Placebo treatment: mean = 16.37; variance = 36.96.

This leads to a t-value of −1.44 with 8 df; the associated **P-value** is 0.19. There is no evidence of a treatment difference.

Response variable Synonymous with **dependent variable**.

Retrospective cohort study See **retrospective study**.

Retrospective study A general term for studies in which all the events of interest occur prior to the onset of the study, and findings are based on looking backward in time. Most common is the *case–control study*, in which comparisons are made between individuals who have a particular condition (the cases) and individuals who do not have the condition (the controls). A sample of cases is selected from the population of individuals who have the problem of interest, and a sample of controls is taken from amongst those individuals known not to have the problem. Information about possible **risk factors** for the condition is then obtained retrospectively for each person in the study by examining past records, by interviewing each person and/or interviewing their relatives, or in some other way. In order to make the cases and controls otherwise comparable, they are frequently matched on characteristics known to be strongly related to both condition and exposure, leading to a *matched case–control study*. Age, sex and socio-economic status are examples of commonly used matching variables. Also commonly encountered is the *retrospective cohort study*, in which a past cohort of individuals is identified from previous information – for example, employment records – and their subsequent mortality or morbidity determined and compared with the corresponding experience of some suitable control group.

Risk A term often used in psychology for the probability that an event will occur over a defined time interval. For example, that the individual will be violent between the ages of 30 and 40 or that a person will become depressed or fail an examination.

Risk factor An aspect of personal behaviour or lifestyle, an environmental exposure, or an inborn or inherited characteristic, which is thought to be associated with a particular disorder.

Robust estimation Methods of estimation that work well, not only under ideal conditions, but also under conditions representing a departure from an assumed distribution or model.

Robust statistics Statistical procedures and tests that still work reasonably well even when the assumptions on which they are based are mildly (or perhaps moderately) violated. **Student's t-test**, for example, is relatively robust against departures from normality.

ROC curves Abbreviation for **receiver operating characteristic curves**. [*Psychological Assessment*, 1997, 9, 171–176].

Rootogram A diagram obtained from a histogram in which the rectangles represent the square roots of the observed frequencies rather than the frequencies themselves. The idea behind such a diagram is to remove the tendency for the variability of a count to increase with its typical size. See also **hanging rootogram**.

Rounding The procedure used for reporting numerical information to fewer decimal places than used during analysis. The rule generally adopted is that excess digits are simply discarded if the first of them is less than five, otherwise the last retained digit is increased by one. So rounding 127.249341 to three decimal places gives 127.249. See also **age heaping** and **digit preference**.

Roy's largest root criterion See **multivariate analysis of variance**.

R-techniques The class of data analysis methods that look for relationships between the variables in a set of **multivariate data**. Includes **principal components analysis** and **factor analysis**. See also **Q-techniques**.

Run-in A period of observation prior to the formation of treatment groups in a **clinical trial** by **random allocation**, during which participants acquire experience with the major components of a study **protocol**. Those participants who experience difficulty complying with the protocol are excluded, while members of the group of proven compliers are randomized into the trial. The rationale behind such a procedure is that, in general, a study with higher compliance will have higher **power** because the observed effects of the difference between treatment groups will not be subjected to the diluting effects of non-compliance.

Runs In a series of observations, the occurrence of an uninterrupted sequence of the same value. For example, in the series 1 1 1 2 2 4 3 3 3 3 3 3, there are four 'runs', the single value, 4, being regarded as a run of length unity. See also **runs test**.

Runs test A test frequently used to detect **serial correlations**. The test consists of counting the number of runs, or sequences of positive and negative residuals, and comparing the result with the expected value under the hypothesis of independence.

Mathematical details

If the sample observations consist of n_1 positive and n_2 negative residuals with both n_1 and n_2 greater than 10, the distribution of the length of runs under independence can be approximated by a **normal distribution** with mean, μ, and variance, σ^2, given by

$$\mu = \frac{2n_1n_2}{n_1 + n_2} + 1$$

$$\sigma^2 = \frac{2n_1n_2(2n_1n_2 - n_1 - n_2)}{(n_1 + n_2)^2(n_1 + n_2 - 1)}$$

See also **Durbin–Watson test**.

S

Sample A selected subset of a population, chosen by some process, usually with the objective of investigating particular properties of the parent population. See also **sampling**.

Sample size The number of individuals to be included in an investigation. Usually chosen so that the study has a particular **power** of detecting an effect of a particular size. Software is available for calculating sample size for many types of study, for example, **nQuery Advisor**. See also **Type II error**.

Sample space The set of all possible outcomes of an experiment. For example, the possible score on tossing a single die is $\{1, 2, 3, 4, 5, 6\}$.

Sample survey A study to estimate particular population characteristics using the information on the characteristics found from a sample of observations taken from the population. See also **random sample**.

Sampling The process of selecting a number of individuals from a population in order to estimate a population parameter or to test a hypothesis about the parameter. To estimate the average IQ of psychology students in the UK, for example, a sample of 100 psychology students might have their IQs assessed. The arithmetic mean of the 100 values would then be used as an estimate of the population mean. Some obvious questions are how to obtain the sample and make the observations and, once the sample data are to hand, how best to use them to estimate the characteristic of the whole population. See also **simple random sampling**.

Sampling distribution The **probability distribution** of a statistic. For example, the sampling distribution of the arithmetic mean of samples of size n, taken from a **normal distribution** with mean μ and standard deviation σ, is also a normal distribution again with mean μ but with standard deviation $\sigma/\sqrt{n}$. More remarkably the same result applies when samples are taken from a non-normal distribution if the sample size is large enough. See the **central limit theorem** entry for some examples.

Sampling error The difference between the sample result and the population characteristic being estimated. In practice, the sampling error can rarely be determined, because the population characteristic is not usually known. With appropriate sampling procedures, however, it can be kept small, and the investigator can determine its probable limits of magnitude. See also **standard error**.

Sampling variation The variation shown by different samples of the same size from the same population.

Sampling with and without replacement Terms used to describe two possible methods of taking samples from a **finite population**. When each element is replaced before the next one is drawn, sampling is said to be 'with replacement'.

When elements are not replaced, then the sampling is referred to as 'without replacement'. See also **bootstrap**, **jackknife** and **hypergeometric distribution**.

Sampling zeros Zero frequencies that occur in the cells of contingency tables simply as a result of inadequate sample size. See also **structural zeros**.

SAS An acronym for Statistical Analysis System, a large computer software system for data processing and data analysis. It contains extensive data management, file handling and graphics facilities, and can be used for most types of statistical analysis, including regression analysis, log-linear models and principal components analysis.

Saturated model A model that contains all main effects and all possible interactions between factors. Since such a model contains the same number of parameters as observations, it results in a perfect fit for a data set. See also **identification**.

Scales of measurement A term used for the different levels of measurement employed in psychological investigations, these ranging from categorical variables through ordinal variables and interval variables to ratio variables.

Scatter Synonym for **dispersion**.

Scatter diagram A two-dimensional plot of a sample of bivariate observations. The diagram is an important aid in assessing what type of relationship links the two variables. An example showing the heights and popularity score for a number of school children is shown in Figure 43. See also **draughtsman's plot**.

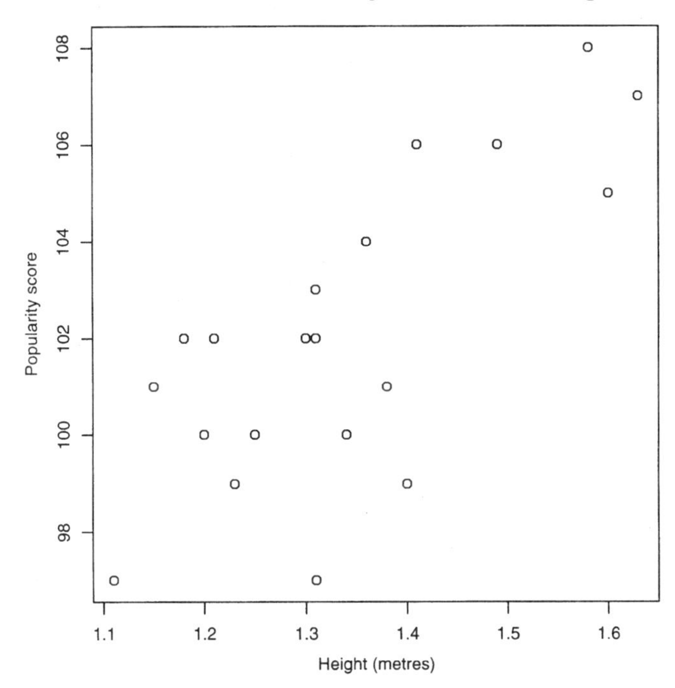

Figure 43 Scatter diagram of heights and popularity score of a sample of school children.

Scattergram Synonym for **scatter diagram**.

Scatterplot Synonym for **scatter diagram**.

Scheffé's test A **multiple comparison test**, which protects against a large **per-experiment error rate**.

Mathematical details

In an **analysis of variance** of a **one-way design**, with g groups, the test uses the following confidence interval for the difference between two means:

$$\bar{y}_{i.} - \bar{y}_{j.} \pm \sqrt{(g-1)F_{g-1,N-g}}\sqrt{\mathrm{MSE}\left(\frac{1}{n_i}+\frac{1}{n_j}\right)}$$

where $\bar{y}_{i.}$ and $\bar{y}_{j.}$ are the observed means of groups i and j, g is the number of groups, MSE is the **error mean square** in the **analysis of variance table**, n_i and n_j are the number of observations in groups i and j, and $F_{g-1,N-g}$ is the F-value for some chosen significance level. The total sample size is represented by N.

Numerical example

Consider an example with $g = 5$ in which the means in the five groups (each with 10 observations) are

A	B	C	D	E
19.67	18.33	27.44	23.44	16.11

and the MSE is 11.83. The Scheffé confidence intervals are as follows:

M1	M2	MD	Confidence interval
A	B	1.33	$1.33 \pm 3.23 \times 1.62 = (-3.90, 6.56)$
A	C	−7.78	$(-13.01, -2.55)^*$
B	D	−3.78	$(-9.01, 1.45)$
A	E	−3.56	$(1.67, 8.79)$
B	C	−9.11	$(-14.34, -3.88)^*$
B	D	−5.11	$(-10.34, 0.12)$
B	E	2.22	$(3.01, 7.45)$
C	D	4.00	$(-1.23, 9.23)$
C	E	11.33	$(6.10, 16.56)^*$
D	E	7.33	$(2.10, 12.56)^*$

* Interval does not include zero, and therefore the corresponding pair of means differ.

Science citation index A database covering 3500 scientific and technical journals in a broad range of disciplines, including psychology. Published by the Institute for Scientific Information (ISI).

Scree plot A plot of the ordered **eigenvalues** of a **correlation matrix**, used to indicate the appropriate number of factors in a **factor analysis** or **principal components analysis**. The critical feature sought in the plot is an 'elbow', the number of factors then being taken as the number of eigenvalues up to this point. Figure 44 shows such a diagram in which the three-component solution is clearly indicated as most appropriate.

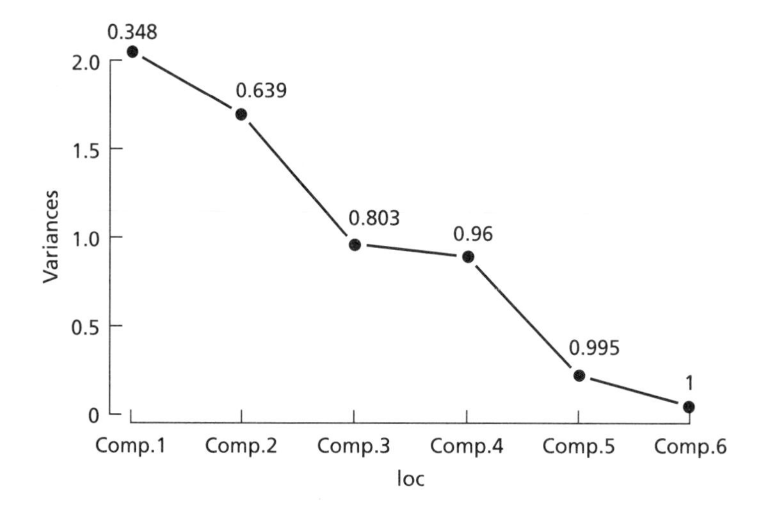

Figure 44 A scree plot.

SD Abbreviation for **standard deviation**.

SE Abbreviation for **standard error**.

Seasonally adjusted A term applied to **time series** from which periodic oscillations with a period of one year have been removed.

Seasonal variation Although strictly used to indicate the cycles in a **time series** that occur yearly, also often used to indicate other periodic movements.

Selection bias The bias that may be introduced into **clinical trials** and other types of psychological investigations, whenever a treatment is chosen by the individual involved or is subject to constraints that go unobserved by the researcher. If there are unobserved factors influencing outcomes and the type of treatment chosen, any direct links between treatment and outcome are confounded with unmeasured variables in the data. A classic example of this problem occurred in the Lanarkshire milk experiment of the 1920s. In this trial, 10 000 children were given free milk supplementation and a similar number received no supplementation. The groups were formed by random allocation. Unfortunately, however, well-intentioned teachers decided that the poorest children should be given priority for free milk, rather than sticking strictly to the original groups. The consequence was that the effects of milk supplementation were indistinguishable from the effects of poverty.

Selection methods in regression A series of methods for selecting 'good' (although not necessarily the best) subsets of explanatory variables when using **regression analysis**. The three most commonly used of these methods are *forward selection*, *backward elimination*, and a combination of both of these known as *stepwise regression*. The criterion used for assessing whether or not a variable should be added to an existing model in forward selection or removed from an existing model in backward elimination is, essentially, the change in the **residual sum of squares** produced by the inclusion or exclusion of the variable.

Mathematical details

Specifically, in forward selection, an 'F-statistic' known as the *F-to-enter*, is calculated as

$$F = \frac{RSS_m - RSS_{m+1}}{RSS_{m+1}/(n + m + 2)}$$

and compared with a pre-set term; calculated Fs greater than the pre-set value lead to the variable under consideration being added to the model. (RSS_m and RSS_{m+1} are the residual sums of squares when models with m and $m+1$ explanatory variables have been fitted.) In backward selection, a calculated F less than a corresponding *F-to-remove* leads to a variable being removed from the current model. In the stepwise procedure, variables are entered as with forward selection, but after each addition of a new variable, those variables currently in the model are considered for removal by the backward elimination process. In this way it is possible that variables included at some earlier stage might later be removed, because the presence of new variables has made their contribution to the regression model no longer important. It should be stressed that none of these automatic procedures for selecting variables is foolproof and they must be used with care.

Numerical example

The following results are for stepwise regression applied to the data set given in the **multiple regression** entry.

Step 0:

Variables in equation			Variables not in equation	
Variable	**Coeff**	***F* to remove**	**Variable**	***F* to enter**
			Years of education	6.19
			Years in job	6.60
			IQ	8.65

Step 1: IQ entered

Variables in equation			Variables not in equation	
Variable	**Coeff**	***F* to remove**	**Variable**	***F* to enter**
IQ	574.03	8.65	Years of education	0.02
			Years in job	22.38

Step 2: Years in job entered

Variables in equation			Variables not in equation	
Variable	**Coeff**	***F* to remove**	**Variable**	***F* to enter**
Years in job	1517.9	22.38	Years of education	6.85
IQ	526.9	26.52		

Step 3: Years of education entered

Variables in equation			Variables not in equation	
Variable	**Coeff**	***F* to remove**	**Variable**	***F* to enter**
Years of education	3380.0	6.85		
Years in job	1814.4	47.76		
IQ	−57.68	0.06		

Step 4: IQ removed

Variables in equation			Variables not in equation	
Variable	**Coeff**	***F* to remove**	**Variable**	***F* to enter**
Years of education	3084.2	64.08	IQ	0.06
Years in job	1786.3	66.10		

Here the final regression equation contains only years in job and years of education.

See also **all subset regression**.

Semi-interquartile range Half the difference between the upper and lower quartiles.

Sensitivity An index of the performance of a diagnostic test, calculated as the percentage of individuals with a disorder who are correctly classified as having the disorder, i.e. the **conditional probability** of having a positive test result given having the disease. A test is sensitive to the disease if it is positive for most individuals having the disease. See also **specificity**, **ROC curves** and **Bayes' Theorem**.

Sequential sum of squares A term encountered primarily in **regression analysis** for the contributions of variables as they are added to the model in a particular sequence. Essentially, the difference in the **residual sum of squares** before and after adding an explanatory variable to the model.

Serial correlation Synonym for **autocorrelation**.

Significance level The level of probability at which it is agreed that the null hypothesis will be rejected. Conventionally set at 0.05.

Significance test A statistical procedure that, when applied to a set of observations, results in a P-value relative to some hypothesis. Examples include **Student's t-test**, **z-test** and **Wilcoxon's signed rank test**.

Sign test A test of the null hypothesis that positive and negative values amongst a series of observations are equally likely. The observations are often differences between a response variable observed under two conditions on a set of subjects. An example of a **distribution-free test**.

Numerical example

A professional body conducts a qualifying examination for all who wish to become members; this is a national examination with some 3000 entrants. In one year the examiners reported that the median mark for all candidates was 54.

The marks of 17 candidates from one particular college were:

38	29	58	41	82	51	45	39	60	42	36	55	46	61	43	52	64

Is there any evidence of a performance below the national average?

Here we take the population median as 54. If the given observations were a random sample from this population then there is a probability of one half that any candidate's mark will be less than the median and a probability of one half that it will be greater. Denoting a mark above 54 by a plus and one below by a minus we have:

−	−	+	−	+	−	−	−	+	−	−	+	−	+	−	−	+

A total of 11 minus and 6 plus. To find a **P-value** we need to calculate the probability of 6 or fewer plus signs in 17 observations in a binomial distribution with $p = \frac{1}{2}$. This can be calculated to be 0.332. There is no evidence that the candidates from this particular college are below the national average.

Similarity coefficient Coefficients, ranging usually from zero to unity, used to measure the similarity of the variable values of two observations from a set of multivariate data. Most commonly used on binary variables. Examples of such coefficients are the matching coefficient and Jaccard's coefficient.

Similarity matrix A symmetric matrix in which values on the main diagonal are unity, and off-diagonal elements are the values of some similarity coefficient for the corresponding pair of individuals.

Mathematical details

A general similarity matrix for n stimuli is of the form

$$\mathbf{S} = \begin{bmatrix} 1.0 & s_{12} & \cdots & s_{1n} \\ s_{21} & 1.0 & \cdots & s_{2n} \\ \vdots & \vdots & \vdots & \vdots \\ s_{n1} & \cdots & \cdots & 1.0 \end{bmatrix}$$

Simple random sampling A form of sampling design in which n distinct units are selected from the N units in the population in such a way that every possible combination of n units is equally likely to be the sample selected. With this type of sampling design the probability that the ith population unit is included in the sample is $\pi_i = n/N$, so that the inclusion probability is the same for each unit. Designs other than this one may also give each unit equal probability of being included, but only here does each possible sample of n units have the same probability.

Simple structure See **factor analysis**.

Simpson's paradox The phenomenon that occurs when the association seen between two variables alters in degree or direction, or both, when a further variable is considered.

Numerical example

The data in the 2×2 contingency table below show the number of white and black defendants found guilty of murder who were given the death penalty. We see that 11.9% of white defendants receive the death penalty compared with 10.2% of black defendants.

	Death penalty	**Not death penalty**
White defendant	19	141
Black defendant	17	149

However, rearranging the overall table into two tables with a further category, colour of victim, gives a totally different picture.

	White victim		Black victim	
	Death	**Not death**	**Death**	**Not death**
White defendant	19	132	0	9
Black defendant	11	52	6	97

In cases with white victims 12.6% of white defendants and 17.5% of black defendants receive the death penalty. In cases with black victims, the corresponding figures are 0% and 5.8%. The conclusion of discrimination against white defendants suggested by the overall table is invalidated by the third variable – colour of victim; opposite conclusions are obtained in each of the two subtables.

Simulation The artificial generation of random processes (usually by means of **pseudorandom numbers** and/or computers) to imitate the behaviour of particular statistical models. See also **Monte Carlo methods**.

Simultaneous confidence interval A **confidence interval** (perhaps more correctly a region) for several parameters being estimated simultaneously.

Single case study Synonym for ***N* of 1 clinical trial**. Sometimes used to study the patterning of neuropsychological performance within a single case in detail. Reports in the published literature may include a series of single case trials or detailed individual accounts whose data are used as information on the replicability of the individual results.

Single linkage clustering A method of **cluster analysis** in which the distance between two clusters is defined as the least distance between a pair of individuals, one member of the pair being in each group.

Numerical example

As an example of the operation of single linkage, the method will be applied to the following distance matrix:

$$\mathbf{D}_1 = \begin{array}{c} \\ 1 \\ 2 \\ 3 \\ 4 \\ 5 \end{array} \begin{array}{c} \begin{array}{ccccc} 1 & 2 & 3 & 4 & 5 \end{array} \\ \begin{pmatrix} 0.0 & & & & \\ 2.0 & 0.0 & & & \\ 6.0 & 5.0 & 0.0 & & \\ 10.0 & 9.0 & 4.0 & 0.0 & \\ 9.0 & 8.0 & 5.0 & 3.0 & 0.0 \end{pmatrix} \end{array}$$

The smallest entry in the matrix is that for individuals 1 and 2; consequently these are joined to form a two-member cluster. Distances between this cluster and the other three individuals are obtained as

$$d_{(12)3} = \min[d_{13}, d_{23}] = d_{23} = 5.0$$

$$d_{(12)4} = \min[d_{14}, d_{24}] = d_{24} = 9.0$$

$$d_{(12)5} = \min[d_{15}, d_{25}] = d_{25} = 8.0$$

A new matrix may now be constructed whose entries are inter-individual distances *and* cluster-individual values:

$$\mathbf{D}_2 = \begin{matrix} (12) \\ 3 \\ 4 \\ 5 \end{matrix} \overset{\begin{matrix} (12) & 3 & 4 & 5 \end{matrix}}{\begin{pmatrix} 0.0 & & & \\ 5.0 & 0.0 & & \\ 9.0 & 4.0 & 0.0 & \\ 8.0 & 5.0 & 3.0 & 0.0 \end{pmatrix}}$$

The smallest entry $\mathbf{D}_2$ is that for individuals 4 and 5; so these now form a second two-member cluster, and a new set of distances is found:

$$d_{(12)3} = 5.0 \text{ as before}$$

$$d_{(12)(45)} - \min[d_{14}, d_{15}, d_{24}, d_{25}] = d_{25} = 8.0$$

$$d_{(45)3} = \min[d_{34}, d_{35}] = d_{34} = 4.0$$

These may be arranged in a matrix $\mathbf{D}_3$:

$$\mathbf{D}_3 = \begin{matrix} (12) \\ 3 \\ (45) \end{matrix} \overset{\begin{matrix} (12) & 3 & (45) \end{matrix}}{\begin{pmatrix} 0.0 & & \\ 5.0 & 0.0 & \\ 8.0 & 4.0 & 0.0 \end{pmatrix}}$$

The smallest entry is now $d_{(45)3}$ and so individual 3 is added to the cluster containing individuals 4 and 5. Finally the groups containing individuals 1, 2 and 3, 4, 5 are combined into a single cluster. The partitions produced at each stage are as follows:

Stage	Groups
P_5	[1], [2], [3], [4], [5]
P_4	[1 2], [3], [4], [5]
P_3	[1 2], [3], [4 5]
P_2	[1 2], [3 4 5]
P_1	[1 2 3 4 5]

The resulting dendrogram is shown in Figure 19 (p. 52).

Single sample *t*-test See **Student's *t*-tests.**

Skewness The lack of symmetry in a **probability distribution**. A distribution is said to have *positive skewness* when it has a long thin tail at the right, and to have *negative skewness* when it has a long thin tail to the left. Figure 45 illustrates both situations.

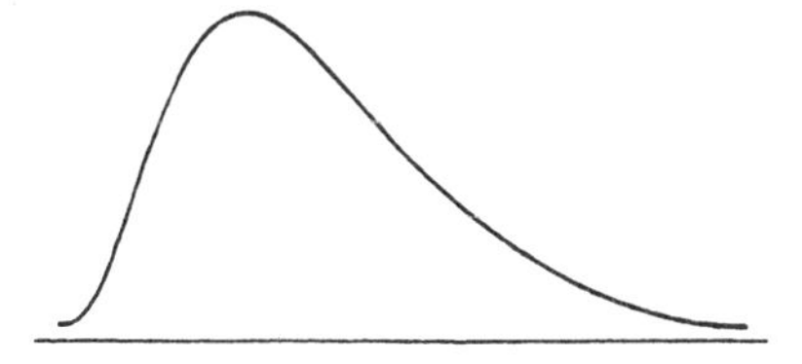
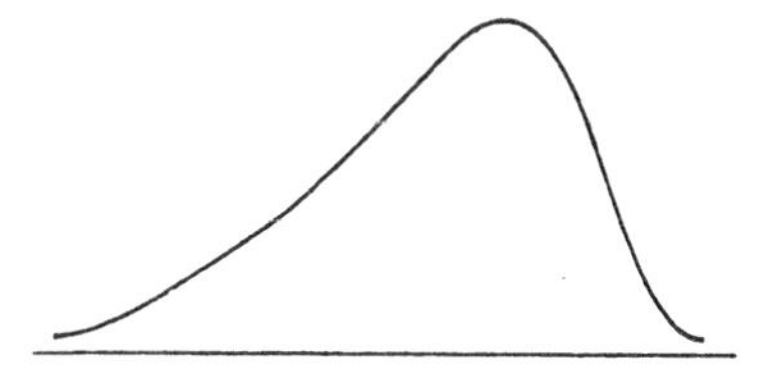

Figure 45 Skewness of distributions: left, positive; right, negative.

Small expected frequencies A term that is found in discussions of the analysis of **contingency tables**. It arises because the derivation of the **chi-squared distribution**, as an approximation for the distribution of the **chi-squared statistic** when the hypothesis of independence is true, is made under the assumption that the expected frequencies are not too small. Typically, this rather vague phrase has been interpreted as meaning that a satisfactory approximation is achieved only when expected frequencies are five or more. Despite the widespread acceptance of this 'rule', it is nowadays thought to be largely irrelevant, since there is a great deal of evidence that the usual chi-squared statistic can be used safely when expected frequencies are far smaller. See also **STATXACT**.

Smoothing A term that could be applied to almost all techniques in statistics that involve fitting some model to a set of observations but which is generally applied to those methods which use computing power to highlight unusual structure very effectively, by taking advantage of people's ability to draw conclusions from well-designed graphics. An example is the construction of a **moving average** for a **time series**.

Social science citation index A database covering 1700 social science journals including psychology. Published by the Institute for Scientific Information (ISI).

Somer's *d* A measure of association for a **contingency table** with ordered row and column categories that is suitable for the asymmetric case in which one variable is considered the response and one explanatory. See also **Kendall's tau statistics**.

Spearman's rho A **rank correlation coefficient** for use when the two variables involved cannot be assumed to have an interval scale.

Mathematical details

If the ranked values of the two variables for a set of n individuals are a_i and b_i, with $d_i = a_i - b_i$, then the coefficient is defined explicitly as

$$r = 1 - \frac{6\sum_{i=1}^{n} d_i^2}{n^3 - n}$$

In essence r is simply **Pearson's product moment correlation coefficient** between the rankings a and b.

Numerical example

The following are the ranks of 10 students based on their marks in French and German

	A	B	C	D	E	F	G	H	I	J
French	1	10	8	5	4	7	6	3	2	9
German	1	10	6	5	7	8	4	2	3	9
d_i	0	0	2	0	3	1	2	1	1	0

$$r = 1 - \frac{6 \times 20}{10 \times 99} = 0.879$$

There is a high degree of correspondence between the results of the tests in French and German.

See also **Kendall's tau statistics**.

Specific variates See **factor analysis**.

Specificity An index of the performance of a diagnostic test, calculated as the percentage of individuals without the disease who are classified as not having the disease, i.e. the **conditional probability** of a negative test result given that the disease is absent. A test is specific if it is positive for only a small percentage of those without the disease. See also **sensitivity**, **ROC curves** and **Bayes' Theorem**.

Sphericity See **Mauchly test**.

Split-half method A procedure used primarily in psychology to estimate the reliability of a test. Two scores are obtained from the same test, either from alternative items, the so-called odd–even technique, or from parallel sections of items. The correlation of these scores, or some transformation of them, gives the required reliability. See also **Cronbach's alpha**. [*Perceptual and Motor Skills*, 1996, 82, 401–402].

S-PLUS A high level programming language with extensive graphical and statistical features that can be used to undertake both standard and non-standard analyses relatively simply.

Spreadsheet In computer technology, a two-way table, with entries which may be numbers or text. Facilities include operations on rows or columns. Entries may also give references to other entries, making possible more complex operations.

SPSS A statistical software package, an acronym for Statistical Package for the Social Sciences. A comprehensive range of statistical procedures is available and, in addition, extensive facilities for file manipulation and recoding or transforming data.

Square contingency table A contingency table with the same number of rows as columns.

Square matrix A matrix with the same number of rows as columns. Correlation matrices and variance–covariance matrices are examples.

Square root transformation A transformation of the form $y = \sqrt{x}$, often used to make random variables suspected to have a Poisson distribution more suitable for techniques such as analysis of variance, by making their variances independent of their means. See also **variance stabilizing transformations**.

Standard deviations (SD) The most commonly used measure of the spread of a set of observations. Equal to the square root of the variance.

Standard error (SE) The standard deviation of the sampling distribution of a statistic. For example, the standard error of the sample mean of n observations is $\sigma/\sqrt{n}$, where σ^2 is the variance of the original observations.

Standardization A term used in a variety of ways in psychological research. The most common usage is in the context of transforming a variable by dividing by its standard deviation to give a new variable with standard deviation unity. This would be known as a *standardized score* or *z-score*. Such scores can be used to compare observations on variables measured on different scales.

Standardized regression coefficient See **beta coefficient**.

Standardized score See **standardization**.

Standard normal distribution A normal distribution with mean zero and variance one.

Standard normal variable A variable having a standard normal distribution.

STATA A comprehensive software package for many forms of statistical analysis; particularly useful for epidemiological and longitudinal data.

STATXACT A specialized statistical package for analysing data from contingency tables that provides exact *P*-values which, in the case of sparse tables, may differ considerably from the values given by, for example, chi-squared tests.

Statistic A numerical characteristic of a sample. For example, the sample mean and sample variance. See also **parameter**.

Stem-and-leaf plot A method of displaying data in which each observation is split into two parts, labelled the 'stem' and the 'leaf': for example, tens and units. The stems are arranged in a column, and the leaves are attached to the relevant stem. The resulting display gives the shape information usually provided by a histogram, whilst retaining the original observation values. Such a plot for the motor cortex neuron interspike intervals in ms for an unstimulated monkey is given in Figure 46.

```
Decimal point is 1 place to the right of the colon

 0 : 26669
 1 : 22344667777888899
 2 : 0002223344455667788888899999
 3 : 00003344455566699
 4 : 11247799
 5 : 0225567899
 6 : 006668
 7 : 23368
 8 : 2
 9 :
10 : 144
```

Figure 46 Stem-and-leaf plot.

Stepwise regression See **selection methods in regression**.

Stochastic process A series of random variables in time or space.

Structural equation modelling A procedure that combines aspects of multiple regression and factor analysis to investigate relationships between latent variables. An example of the use of such a model would be to assess the stability over time of attitudes such as alienation and the relation to background variables such as education and occupation. See also **LISREL** and **EQS**. [*Journal of Applied Psychology*, 1998, 83(3), 471–785].

Structural zeros Zero frequencies occurring in the cells of a contingency table, which arise because it is theoretically impossible for an observation to fall in the cell. For example, if male and female students are asked about health problems that cause them concern, then the cell corresponding to, say, menstrual problems for men will have a zero entry. Special adaptations of log-linear models can be used to analyse such data. See also **sampling zeros**.

Studentized range statistic A statistic that occurs most often in multiple comparison tests.

Mathematical details

Defined as

$$q = \frac{\bar{x}_{\text{largest}} - \bar{x}_{\text{smallest}}}{\sqrt{\text{MSE}/n}}$$

where $\bar{x}_{\text{largest}}$ and $\bar{x}_{\text{smallest}}$ are the largest and smallest means amongst the means of k groups, and MSE is the **error mean square** from an **analysis of variance** of the groups.

Studentized residual See **residual**.

Student's *t*-distribution The **probability distribution** of the ratio of a **standard normal variable** to the square root of a variable with a **chi-squared distribution**. The shape of the distribution varies with n and, as n gets larger, the shape of the t-distribution approaches that of the **standard normal distribution**.

Student's *t*-tests Significance tests for assessing hypotheses about population means. One version is used in situations where it is required to test whether the mean of a population takes a particular value. This is generally known as a *single sample t-test*. Another version is designed to test the equality of the means of two populations. When independent samples are available from each population, the procedure is often known as the *independent samples t-test*.

Mathematical details

The test statistic for an independent samples t-test is defined as

$$t = \frac{\bar{x}_1 - \bar{x}_2}{s\sqrt{\frac{1}{n_1} + \frac{1}{n_2}}}$$

where $\bar{x}_1$ and $\bar{x}_2$ are the means of samples of size n_1 and n_2 taken from each population, and s^2 is an estimate of the assumed common variance, given by

$$s^2 = \frac{(n_1 - 1)s_1^2 + (n_2 - 1)s_2^2}{n_1 + n_2 - 2}$$

If the null hypothesis of the equality of the two population means is true, t has a **Student's *t*-distribution** with $n_1 + n_2 - 2$ degrees of freedom, allowing ***P*-values** to be calculated. The test assumes that each population has a **normal distribution**, but is known to be relatively insensitive to departures from this assumption.

Numerical example

The following data were collected in a study of social pressure, in which two groups of 10 students were asked to estimate the length of a piece of string. In one group, the participant is placed in the presence of four other persons who, unknown to him/her, have been instructed to give estimates which are too large. Each of these persons gives their estimate orally to the experiment before the participant is asked for his/her estimate. In the other groups the same procedure is employed except that the four persons are instructed to make honest estimates.

Estimates made in unbiased environment	Estimates made in biased environment
38	47
35	45
37	42
40	45
45	47
34	40
40	43
36	45
43	40
41	48
$\bar{x}_1 = 38.9, s_1^2 = 11.3$	$\bar{x}_2 = 44.2, s_2^2 = 7.4$

$t = -3.68$ with 18 df. The associated P-value is less than 0.05 and the hypothesis of equality of means is rejected, suggesting that estimates made in the biased environment are themselves biased.

See also **matched pairs *t*-test**.

Supervised pattern recognition See **pattern recognition**.

Survival curve See **survival function**.

Survival function The probability that the **survival time** of an individual is larger than some particular value. A plot of this probability against time is called a *survival curve* and is a useful component in the analysis of such data. See also **product limit estimator**.

Survival time Observations of the time until the occurrence of a particular event: for example, recovery, improvement or death.

Symmetric distribution A **probability distribution** or **frequency distribution** that is symmetrical about some central value. The **normal distribution** is an example.

Symmetric matrix A matrix in which the corresponding row/column, column/row off-diagonal elements are equal. Statistical examples are **correlation matrices** and **variance–covariance matrices**.

Synergism A term used when the joint effect of two treatments is greater than the sum of their effects when administered separately (*positive synergism*) or when the sum of their effects is less than when administered separately (*negative synergism*).

T

Target population The collection of individuals, items, measurements, etc. about which it is required to make inferences. Often the population actually sampled differs from the target population and this may result in misleading conclusions being drawn.

Taylor's power law A convenient method for finding an appropriate transformation of grouped data, to make them satisfy the homogeneity of variance assumption of techniques such as the **analysis of variance**. The method involves calculating the slope of the regression line of the logarithm of the group variances against the logarithm of the group means, i.e. b in the equation

$$\log_{10} s_i^2 = a + b \log_{10} \bar{x}_i$$

The value of $1 - b/2$ indicates the transformation needed, with non-zero values corresponding to a particular transformation involving a particular power, and zero corresponding to a **logarithmic transformation**.

Test statistic A statistic used to assess a particular hypothesis in relation to some population. The essential requirement of such a statistic is a known **probability distribution** when the null hypothesis is true. Examples are found in the **Student's t-test** and the **z-test** entries.

Tetrachoric correlation An estimate of the correlation between two **random variables** having a **bivariate normal distribution**, obtained from the information from a double dichotomy of their bivariate distribution. For example, IQ dichotomized at 100 and reaction time to some task dichotomized at 1 second. Can be estimated by **maximum likelihood estimation**.

Therapeutic trial Synonym for **clinical trial**.

Threshold model A model that postulates that an effect occurs only above some threshold value. For example, a model that assumes that the effect of a psychological treatment is zero if the patient does not attend for at least a certain number of sessions.

Tied observations A term usually applied to ordinal variables to indicate observations that take the same value on a variable.

Time-dependent covariates **Covariates** whose values change over time: for example, age, dose of drug, reinforcement schedule. See also **time-independent covariates**.

Time-independent covariates **Covariates** whose values remain constant over time, for example, gender. See also **time-dependent covariates**.

Time series Values of a variable recorded, usually at a regular interval, over a long period of time. An example involving the pain scores of a person suffering from a migraine headache is shown in Figure 47. The observed movement and fluctuations of many such series are composed of four different components: secular trend, seasonal variation, cyclical variation and irregular variation. Such data usually require special methods for their analysis because of the presence of serial correlations between the separate observations. The most commonly applied methods for such data are autoregressive models, where the aim is generally to be able to make forecasts and predictions.

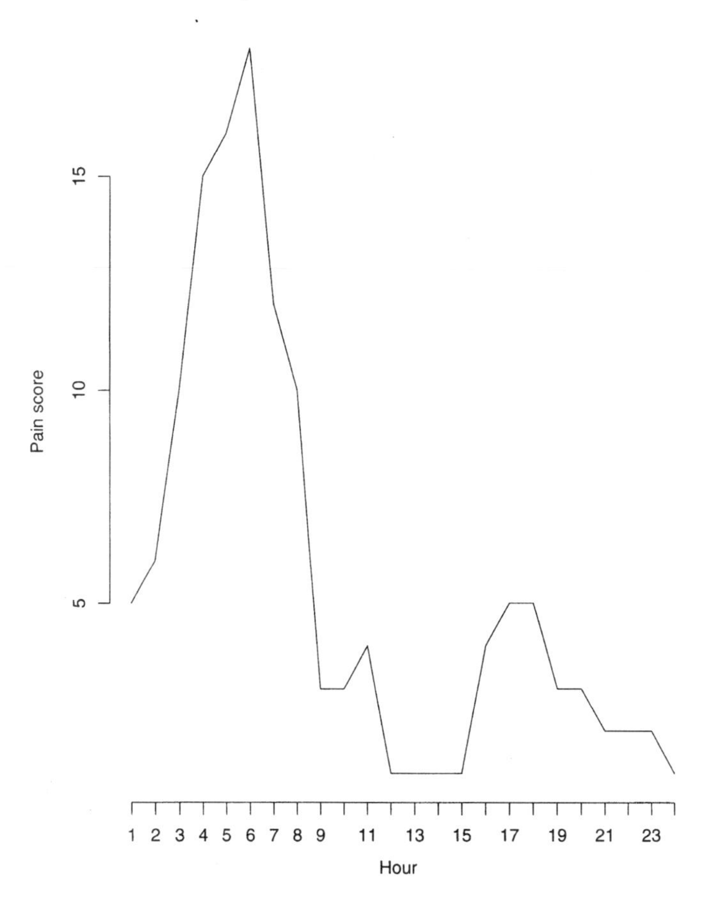

Figure 47 A time series; pain scores over 24 hours for a person suffering from a migraine headache.

Time-varying covariates Synonym for **time-dependent covariates**.

Tolerance A term used in **stepwise regression** for the proportion of the sum of squares about the mean of an explanatory variable, not accounted for by other variables already included in the regression equation. Small values indicate possible **multicollinearity** problems.

Total sum of squares The sum of the squared deviations of all the observations from their mean.

Total sums of squares and cross-products matrix See **multivariate analysis of variance**.

Trace of a matrix The sum of the elements on the main diagonal of a square matrix; usually denoted as tr(**A**). So, for example, if

$$\mathbf{A} = \begin{pmatrix} 3 & 2 \\ 4 & 1 \end{pmatrix}$$

then $\text{tr}(\mathbf{A}) = 4$. Occurs in accounts of techniques for multivariate analysis.

Tracking A term sometimes used in discussions of longitudinal data to describe the ability to predict subsequent observations from earlier values. Informally, this implies that participants who have, for example, the largest values of the dependent variable at the start of the study tend to continue to have the larger values. More formally, a population is said to track with respect to a particular observable characteristic if, for each individual, the expected value of the relevant deviation from the population mean remains unchanged over time.

Transformation A change in the scale of measurement for some variable(s). Examples are the square root transformation and logarithm transformation. Generally used in an attempt to satisfy the assumptions of particular methods, for example normality.

Treatment allocation ratio The ratio of the number of participants allocated to the two treatments in a clinical trial. Equal allocation is most common in practice, but it may be advisable to allocate participants randomly in other ratios when comparing a new treatment with an old, or when one treatment is much more difficult or expensive to administer. The chance of detecting a real difference between the two treatments is not reduced much as long as the ratio is not more extreme than 2:1.

Treatment cross-contamination Any instance in which a participant assigned to receive a particular treatment in a clinical trial is exposed to one of the other treatments during the course of the trial.

Treatment–period interaction Synonym for **carryover effect**.

Treatment trial Synonym for **clinical trial**.

Trend Movement in one direction of the values of a variable over a period of time. Also popularly used by psychologists to refer to results which just fail to reach a conventional level of significance.

Trimmed mean A mean calculated after some proportion of the observations at the extremes of a variable's frequency distribution are removed. Used to reduce the effect of outliers.

Truncated data Data for which sample values larger (truncated on the right) or smaller (truncated on the left) than a fixed value are either not recorded or not observed. See also **censored observations**.

Two-by-two (2 × 2) contingency table A **contingency table** with two rows and two columns formed from cross-classifying two **binary variables**. The general form of such a table is:

	Variable 1	
Variable 2	**0**	**1**
0	a	b
1	c	d

Two-by-two (2 × 2) crossover design See **crossover design**.

Two-dimensional contingency table See **contingency table**.

Two-sided test A test where the alternative hypothesis is not directional: for example, that one population mean is not equal to another. See also **one-sided test**.

Two-way classification The classification of a set of observations according to two criteria, as, for example, in a **contingency table** constructed from two variables.

Type I error The error that results when the null hypothesis is falsely rejected.

Type II error The error that results when the null hypothesis is falsely accepted.

Type III error It has been suggested by a number of authors that this term should be used for identifying the poorer of two therapeutic procedures as the better.

Type I, II and III sums of squares Terms for three types of sums of squares encountered in the analysis of **unbalanced designs**. The first describes an analysis approach in which the independent variables are used to explain the variation of the dependent variable in a particular sequence. The latter refers to the sum of squares of the dependent variable explained by an independent variable when it is last in the sequence. Although Type III sums of squares are often recommended, most statisticians would be unhappy considering sequences in which interaction terms are listed before the main effects from which they are composed. So, for example, in a two factor unbalanced design, looking at the effects in the order AB, B, A to find a Type III sums of squares for A would not be considered appropriate.

U

Unbalanced design See **balanced design**.

Unbiased See **bias**.

Unidentified model See **identification**.

Uniform distribution The probability distribution of a random variable having constant probability over an interval.

Mathematical details

Specifically, the distribution function is given by

$$f(x) = \frac{1}{\beta - \alpha}, \qquad \alpha < x < \beta$$

The mean of the distribution is $(\alpha + \beta)/2$ and the variance is $(\beta - \alpha)^2/12$. The most commonly encountered uniform distribution is one in which the parameters α and β take the values 0 and 1, respectively.

Unimodal distribution A probability distribution or frequency distribution having only a single mode. See the normal distribution entry for an example.

Unit normal variable Synonym for **standard normal variable**.

Univariate contrast analysis Synonym for **planned comparison**.

Univariate data Data involving a single measurement on each subject or patient.

Universe A little-used alternative term for **population**.

Unsupervised pattern recognition See **pattern recognition**.

Unweighted means analysis An approach to the analysis of two-way and higher order factorial designs when there are an unequal number of observations in each cell. The analysis is based on cell means, using the harmonic mean of all cell frequencies as the sample size for all cells.

U-shaped distribution A probability distribution or frequency distribution shaped more or less like a letter U, though not necessarily symmetrical.

Such a distribution has its greatest frequencies at the two extremes of the range of the variable. An example is shown in Figure 48.

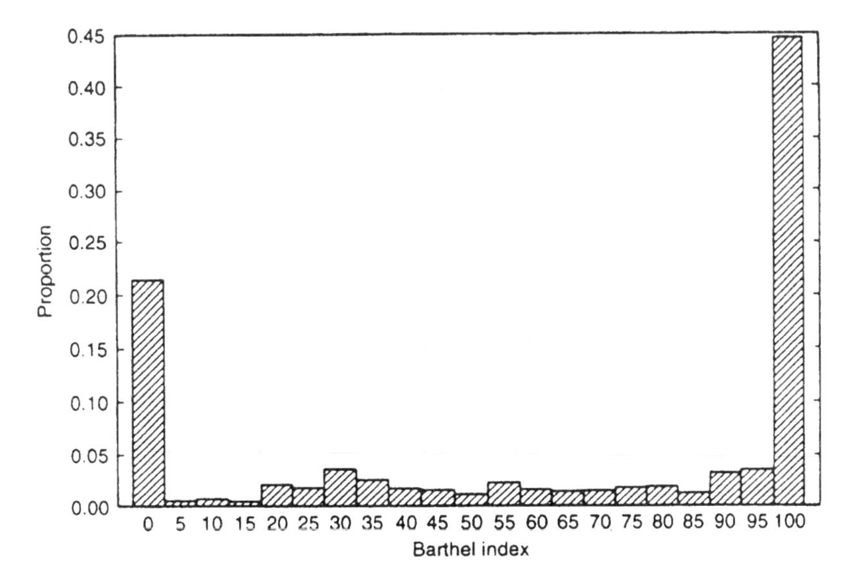

Figure 48 U-shaped distribution.

V

Validity The extent to which a measuring instrument is measuring what was intended.

Validity checks A part of **data editing** in which a check is made that only allowable values or codes are given for the answers to questions asked of subjects. A negative value for a participant's height, for example, would clearly not be an allowable value.

Variable Some characteristic that differs from subject to subject or from time to time.

Variable selection The problem of selecting subsets of variables, in **regression analysis**, that contain most of the relevant information in the full data set. See also **adequate subset**, **all subsets regression** and **selection methods in regression**.

Variance A measure of the spread of a set of observations.

Mathematical details

In a population the variance σ^2 is defined as

$$\sigma^2 = E(x - \mu)^2$$

where E denotes **expected value** and μ is the population mean. An unbiased estimator of the population value is provided by s^2, given by

$$s^2 = \frac{1}{n-1}\sum_{i=1}^{n}(x_i - \bar{x})^2$$

where $x_1, x_2, \ldots, x_n$ are the n sample observations and $\bar{x}$ is the sample mean.

Numerical example

Consider the following set of 10 IQ scores

$$101, 105, 98, 91, 115, 88, 102, 103, 99, 107$$

The mean is 100.9.

The variance is

$$\begin{aligned}&\tfrac{1}{9}[(101 - 100.9)^2 + (105 - 100.9)^2 + (98 - 100.9)^2 + (91 - 100.9)^2\\&\quad + (115 - 100.9)^2 + (88 - 100.9)^2 + (102 - 100.9)^2\\&\quad + (103 - 100.9)^2 + (99 - 100.9)^2 + (107 - 100.9)^2] = 59.43\end{aligned}$$

Variance components Variances of random effect terms in **linear models**. For example, in a simple **mixed effects model** for **longitudinal data**, both subject effects and error terms are random, and estimation of their variances is of some importance. In the case of a **balanced design**, estimation of these variances is usually achieved directly from the appropriate analysis of variance table by equating **mean squares** to their **expected values**. When the data are unbalanced, a variety of estimation methods might be used, with **maximum likelihood estimation** the most common. [*Developmental Psychology*, 1998, 34, 125–129].

Variance–covariance matrix A **symmetric matrix** in which the off-diagonal elements are the **covariances** (sample or population) of pairs of variables, and the elements on the main diagonal are the variances (sample or population) of the variables.

Mathematical details

The population variance–covariance matrix is generally denoted by Σ, a matrix having the form

$$\Sigma = \begin{bmatrix} \sigma_1^2 & \sigma_{12} & \cdots & \sigma_{1q} \\ \sigma_{21} & \cdots & \cdots & \cdots \\ \vdots & \vdots & \vdots & \vdots \\ \sigma_{q1} & \cdots & \cdots & \sigma_q^2 \end{bmatrix}$$

where σ_i^2 is the variance of variable i and σ_{ij} is the covariance of variables i and j.

Numerical example

The data below are the ratings of a particular type of behaviour for 15 children made by a qualified psychologist and by three psychology students.

Psychologist	Student 1	Student 2	Student 3
6.3	5.0	4.8	6.0
4.1	3.2	3.1	3.5
5.1	3.6	3.8	4.5
5.0	4.5	4.1	4.3
5.7	4.0	5.2	5.0
3.3	2.5	2.8	2.6
1.3	1.7	1.4	1.6
5.8	4.8	4.2	5.5
2.8	2.4	2.0	2.1
6.7	5.2	5.3	6.0
1.5	1.2	1.1	1.2
2.1	1.8	1.6	1.8
4.6	3.4	4.1	3.9
7.6	6.0	6.3	6.5
2.5	2.2	1.6	2.0

The lower triangular part of the sample covariance matrix is

Psychologist	3.88			
Student 1	2.81	2.12		
Student 2	3.15	2.27	2.66	
Student 3	2.51	2.57	2.83	3.24

Variance inflation factor An indicator of the effect the other explanatory variables have on the variance of a **regression coefficient** of a particular variable. Given by the reciprocal of the square of the **multiple correlation coefficient** of the variable with the remaining variables. Indicates the strength of the linear relationship between a variable and the remaining explanatory variables. A rough rule of thumb is that values greater than 10 give some cause for concern.

Mathematical details

The variance inflation factor VIF_j for the jth variable is given by

$$\text{VIF}_j = \frac{1}{1 - R_j^2}$$

where R_j^2 is the square of the multiple correlation from the regression of the jth explanatory variable on the remaining explanatory variables.

Variance ratio distribution Synonym for ***F*-distribution**.

Variance ratio test Synonym for ***F*-test.**

Variance stabilizing transformations Transformations designed to give approximate independence between mean and variance as a preliminary to, for example, **analysis of variance**. The **square root transformation** for count data is an example.

Varimax rotation A method for **factor rotation** that, by maximizing a particular function of the initial **factor loadings**, attempts to find a set of factors satisfying, approximately at least, a **simple structure**.

Vector A matrix having only one row or column.

Venn diagram A graphical representation of the extent to which two or more quantities or concepts are mutually inclusive and mutually exclusive. An example is given in Figure 49.

Visual analogue scales Scales used to measure quantities such as pain or satisfaction. The patient is shown a straight line, the ends of which are labelled with extreme status. They are then asked to mark the point on the line, which represents their perception of their current state. For example, such a scale for worry might be:

no worry ---------------------------------- constant worry

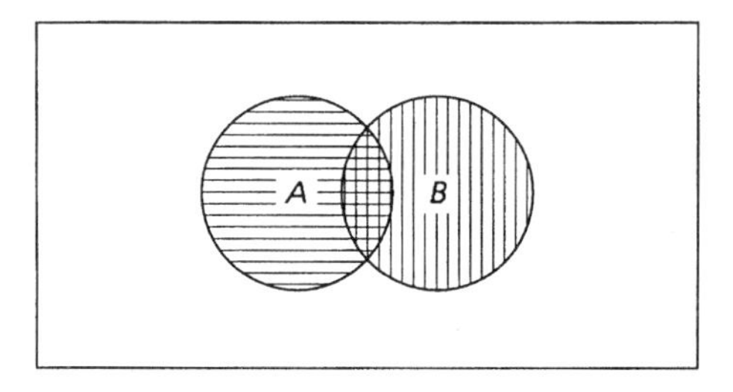

Figure 49 Venn diagram.

Volunteer bias A possible source of **bias in clinical trials** involving volunteers, but not involving random allocation, because of the known propensity of volunteers to respond better than other participants. [*Archives of Sexual Behaviour*, 1998, 27, 181–195].

Ward's method An agglomerative hierarchical clustering method in which a sum-of-squares criterion is used to decide on which individuals or which clusters should be fused at each stage in the procedure. See also **single linkage, average linkage, complete linkage** and **K-means cluster analysis**. [*Neuropsychology*, 1996, 10, 66–73].

Ward's minimum variance procedure Synonym for **Ward's method**.

Wash-out period An interval introduced between the treatment periods in a crossover design in an effort to eliminate possible carryover effects.

Weighted average An average of quantities which have an attached series of weights in order to make proper allowance for their relative importance. For example, a weighted arithmetic mean of a set of observations, $x_1, x_2, \ldots, x_n$, with weights $w_1, w_2, \ldots, w_n$ is given by

$$\frac{\sum_{i=1}^{n} w_i x_i}{\sum_{i=1}^{n} w_i}$$

Weighted kappa A version of the kappa coefficient that permits disagreements between raters to be differentially weighted, to allow for differences in how serious such disagreements are judged to be.

Mathematical details

We use the same nomenclature for the observed ratings as in the kappa statistic entry but now introduce weights w_{ij} with $w_{ii} = 1$ and the other values reflecting a judgement as to the seriousness of the disagreements and having values between 0 and 1 so that all disagreements are given less weight than exact agreement. The weighted kappa statistic is then

$$K^{(w)} = \frac{P_0^{(w)} - P_c^{(w)}}{1 - P_c^{(w)}}$$

$$\text{where} \quad P_0^{(w)} = \frac{1}{N} \sum_{i=1}^{c} \sum_{j=1}^{c} w_{ij} n_{ij}$$

$$P_c^{(w)} = \frac{1}{N^2} \sum_{i-1}^{c} \sum_{j=1}^{c} w_{ij} n_{i.} n_{.j}$$

Numerical example

Using the data under the **kappa statistic** entry and assigning weights as follows:

	1	2	3
1	1.0	0.5	0.0
2	0.5	1.0	0.5
3	0.0	0.5	1.0

$$P_0^{(w)} = \frac{1}{100} \times [24 + 0.5 \times 13 + 0 \times 3 + 0.5 \times 5 + 20 + 0.5 \times 5 + 0 \times 1 + 0.5 \times 7 + 22] = 0.71$$

$$P_c^{(w)} = \frac{1}{100^2}\begin{bmatrix} 1 \times 40 \times 30 + 0.5 \times 40 \times 40 + 0 \times 40 \times 30 \\ +0.5 \times 30 \times 30 + 1 \times 30 \times 40 + 0.5 \times 30 \times 30 \\ +0 \times 30 \times 30 + 0.5 \times 30 \times 40 + 1 \times 30 \times 30 \end{bmatrix} = 0.56$$

$$K^{(w)} = \frac{0.71 - 0.56}{1 - 0.56} = 0.34$$

The unweighted kappa value for this example is 0.49. The weighted version is reduced because of those disagreements which are considered particularly poor.

[*Journal of Autism and Developmental Disorders*, 1998, 28, 69–75].

Weighted least squares A method of estimation in which estimates arise from minimizing a weighted sum of squares of the differences between the response variable and its predicted value in terms of the model of interest. Often used when the variance of the response variable is thought to change over the range of values of the explanatory variable(s), in which case the weights are generally taken as the reciprocals of the variance. See also **least squares** and **iteratively weighted least squares**.

White noise sequence A sequence of independent **random variables** that all have a **normal distribution** with zero mean and the same variance. An example is shown in Figure 50.

Wilcoxon's rank sum test An alternative name for the **Mann–Whitney test**.

Wilcoxon's signed rank test A **distribution-free method** for testing the difference between two populations using matched samples. The test is based on the absolute differences of the pairs of observations in the two samples, ranked according to size, with each rank being given the sign of the original difference. The **test statistic** is the sum of the positive ranks.

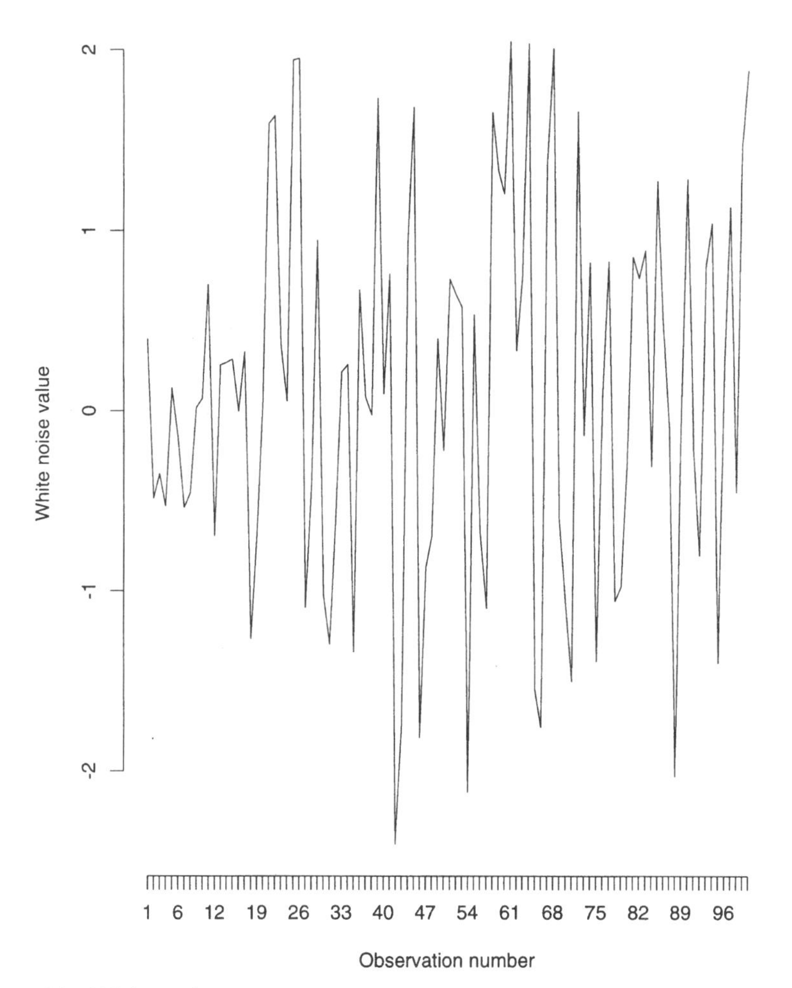

Figure 50 White noise sequence.

Wilks' lambda See **multivariate analysis of variance**.

Within groups mean square See **analysis of variance**.

Within groups sum of squares See **analysis of variance**.

Within groups sums of squares and cross-products matrix See **multivariate analysis of variance**.

Y

Yates' correction When testing for independence in a **contingency table**, a continuous **probability distribution**, namely the **chi-squared distribution**, is used as an approximation to the discrete probability of observed frequencies, namely the **multinomial distribution**. To improve this approximation, Yates suggested a correction that involves subtracting 0.5 from the positive discrepancies (observed – expected) and adding 0.5 to the negative discrepancies before these values are squared in the calculation of the usual **chi-squared statistic**. If the sample size is large, the correction will have little effect on the value of the test statistic.

Mathematical details

Suppose we are interested in the following 2×2 contingency table:

	Variable A		
Variable B	**Category 1**		**Category 2**
Category 1	a	b	$a+b$
Category 2	c	d	$c+d$
	$a+c$	$b+d$	$N=a+b+c+d$

The usual chi-squared statistics can be written as

$$\chi^2 = \frac{N(ad-bc)^2}{(a+b)(c+d)(a+c)(b+d)}$$

With Yates' correction the statistic becomes

$$\chi^2 = \frac{N(|ad-bc|-0.5N)^2}{(a+b)(c+d)(a+c)(b+d)}$$

where $|ad-bc|$ denotes the absolute value of $ad-bc$.

Numerical example

The following data give the number of phobic patients who improve in a clinical trial comparing a new drug with a placebo:

Treatment	**Improve**	**Do not improve**	**Total**
Drug	15	35	50
Placebo	4	46	50
Total	19	81	100

The usual chi-squared statistic is

$$\chi^2 = \frac{100 \times (15 \times 46 - 35 \times 4)^2}{50 \times 50 \times 19 \times 81} = 7.86$$

with Yates' correction this becomes

$$\chi^2 = \frac{100 \times (550 - 50)^2}{50 \times 50 \times 19 \times 81} = 6.50$$

With 1 df this value is highly significant.

Z

z-scores See **standardization**. [*Perceptual and Motor Skills*, 1997, 84, 1284–1286].

z-test A test for assessing hypotheses about population means when the population variances are known. For example, for testing that the means of two populations having **normal distributions** are equal.

Mathematical details

To test the hypothesis H_0: $\mu_1 = \mu_2$, when the variance of each population is known to be σ^2, the **test statistic** is

$$z = \frac{\bar{x}_1 - \bar{x}_2}{\sigma\sqrt{\frac{1}{n_1} + \frac{1}{n_2}}}$$

where $\bar{x}_1$ and $\bar{x}_2$ are the means of samples of size n_1 and n_2 from the two populations. If H_0 is true, z has a **standard normal distribution**.

Numerical example

The IQs of a random sample of 10 cricket lovers and 10 football lovers were:

Cricket	Football	Cricket	Football
106	97	108	96
110	90	99	110
115	125	97	93
121	95	101	100
95	101	98	95

Is there any evidence of a difference in the mean IQs of the two populations? We can assume the population standard deviation is 15.

$$z = \frac{105 - 100.2}{15\sqrt{\frac{1}{10} + \frac{1}{10}}} = 0.72$$

There is no evidence of a difference.

See also **Student's *t*-tests**.

z transformation See **Fisher's z transformation**. [*Journal of Psychology*, 1997, 13, 572–574].